Heat Exchangers - Basics Design Applications

Heat Exchangers - Basics Design Applications

Editor

Mergen Balik

scitus
academics

Heat Exchangers - Basics Design Applications
Edited by **Mergen Balik**

ISBN: 978-1-68117-490-7
Library of Congress Control Number: 2016942921

www.scitusacademics.com
Box No. 4766, 616 Corporate Way,
Suite 2, Valley Cottage,
NY 10989

Notice

Printed in United States of America on Acid Free Paper

Preface

A heat exchanger is a device designed to transfer or possibly exchange heat from one material to the next effectively. When a fluid is used to transport heat, it may be a liquid, such as oil or water, or it may be air moving through a substance. They are popular in heating spaces, natural gas processing, oil refineries, petrochemical plants, chemical plants, power plants, air conditioning, cooling and sewage treatment. A motor radiator is probably the most recognized type of heat exchanger. In a radiator, a formula of water and ethylene glycol, commonly known as antifreeze, transfers heat from the engine to the radiator, after which it flows from the radiator into the surrounding air. This procedure helps you to overheat the engine of a car. For efficiency, heat exchangers are intended to capitalize on the surface part of the wall between the two fluids while reducing the resistance of the exchanger to substance flow. The performance of the exchanger can also be affected by the addition of corrugations or fins in one or both directions, which increase the surface area as well as canalize the fluid flow or possibly cause turbulences.

This book provides extensive information on specific heat exchanger cases. In addition to the issues of thermodynamic fundamentals, the book addresses many important issues, such as conceptions, design, operations, fouling and cleaning of heat exchangers. The book is not inevitably expected to be a basic source of knowledge in the area covered, but a moderate guide to detailed solutions to specific technical problems faced by engineers and technicians engaged in research and development in the fields of heat transfer and heat exchangers.

Table of Contents

CHAPTER 1

Design of earth–air heat exchanger system

Trilok Singh Bisoniya

Department of Mechanical Engineering, Sagar Institute of Research & Technology-Excellence, Bhopal 462041, India

ABSTRACT

The earth–air heat exchanger (EAHE) is a promising technique which can effectively be used to reduce the heating/cooling load of a building by preheating the air in winter and vice versa in summer. In the last two decades, a lot of research has been done to develop analytical and numerical models for the analysis of EAHE systems. Many researchers have developed sophisticated equations and procedures but they cannot be easily recast into design equations and must be used by trial-and-error. In this paper, the author has developed a one-dimensional model of the EAHE systems using a set of simplified design equations. The method to calculate the earth's undisturbed temperature (EUT) and more recently developed correlations for friction factor and Nusselt number are used to ensure higher accuracy in the calculation of heat transfer. The developed equations enable designers to calculate heat transfer, convective heat transfer coefficient, pressure drop, and length of pipe of the EAHE system. A longer pipe of smaller diameter buried at a greater depth and having lower air flow velocity results in increase in performance of the EAHE system.

KEYWORDS

Earth–air heat exchanger, Nusselt number (N_u), Effectiveness of EAHE (ε), Earth's undisturbed temperature (EUT)

1. BACKGROUND

The consumption of high-grade energy has increased considerably with growing needs to achieve thermal comfort conditions inside buildings, residential, greenhouses, livestock buildings, etc. It is desirable to minimize the use of high-grade energy consumption and to promote the use of renewable energy in order to save the earth from hazardous effects of global warming and ozone layer depletion. Numerous alternative techniques are being currently explored to achieve thermal comfort conditions inside buildings. The earth–air heat exchanger is one of these promising techniques which can effectively be used to preheat the air in winter and vice versa in summer. The temperature of earth at a depth of 1.5 to 2 m remains fairly constant throughout the year (Bisoniya et al. 2013). This constant temperature is called earth's undisturbed temperature (EUT). The EUT remains higher than ambient air temperature in winter and lower than ambient air temperature in summer. The concept of earth–air heat exchanger (EAHE) is very simple as shown in Fig. 1. The ambient air is drawn through the pipes of the EAHE buried at a particular depth, moderated to EUT, and gets heated in winter and vice versa in summer. In this way, the heating and cooling load of building can be reduced passively.

In the last two decades, a lot of research has been done to develop analytical and numerical models for analysis of the EAHE systems (Mihalakakou et al. 1994; Bojic et al. 1997; Gauthier et al. 1997; Hollmuller and Lachal 2001; Su et al. 2012; Sehli et al.2012; Ozgener et al. 2013). The performance analysis of EAHE involved either the calculation of conductive heat transfer from the pipe to the ground mass or the calculation of convective heat transfer from the circulating air to the pipe and changes in the air temperature and humidity. A number of computer modeling tools are commercially available. EnergyPlus and TRNSYS have EAHE modules that work well; however, these are analysis tools and are not quickly used for design.

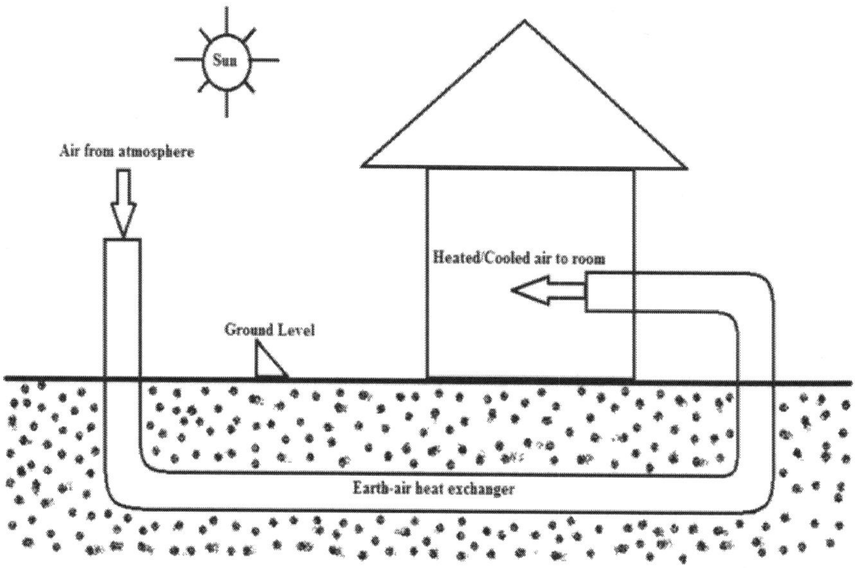

Figure 1. One-dimensional model of earth–air heat exchanger system

Presently, Computational Fluid Dynamics (CFD) is very popular among researchers for modeling and performance analysis of the EAHE systems. The CFD employs a very simple rule of discretization of whole system in small grids. Then, governing equations were applied on these discrete elements to get numerical solutions concerning flow parameters, pressure distribution, and temperature gradients in less time and at reasonable cost because of reduced required experimental work (Kanaris et al. 2006; Wang et al. 2007). For complete analysis of an EAHE system, the use of CFD is recommended, but it is limited to those who have a good command over it. For the initial design of an EAHE system, the use of basic heat transfer equations is more suitable to determine the geometrical dimensions of the system. Many researchers like De Paepe and Janssens (2003), Badescu and Isvoranu (2011), and T'Joen et al. (2012) have developed EAHE design equations and procedures.

In this paper, the author has developed a one-dimensional model of the EAHE system. The method to calculate the EUT and the more recently developed correlations for friction factor and Nusselt number are used to ensure higher accuracy in calculation of heat transfer.

1.1 Analysis and modeling of EAHE

The development of the model of the EAHE system involves the use of basic heat transfer equations. The geometrical dimensions of the EAHE system are decided by taking into account the amount of heating or cooling load to be met for space conditioning of the building. The design procedure includes identifying the input parameters which are known to the user and the parameters affecting desired design output. Once the design output is fixed, the heat transfer equations are manipulated to meet the desired output in terms of input parameters.

Therefore, depending on the nature of the design problem the mass flow rate of air, $\dot{m}$; inlet air temperature, T_{in}; desired outlet air temperature from EAHE, T_{out}; and EUT are considered as parameters of the sizing problem. Furthermore, it is considered that the location of installation of the EAHE system is known. So, ambient air temperature and soil properties are known. The EUT temperature is estimated as the annual average ambient air temperature of a particular location; therefore, it is also assumed as a known parameter. The mass flow rate of air and the outlet air temperature are set by the design requirements.

The geometric sizing parameters of an EAHE include the diameter of the pipe, D; length of the pipe, L; and number of pipes in parallel, N_p, in the heat exchanger.

1.2 Assumptions

The following assumptions were made in order to simplify the development of the one-dimensional model of the EAHE system:

- The surface temperature of the ground is defined as equal to the ambient air temperature, which equals the inlet air temperature.
- EUT can bė approximated to the annual average temperature of the location (Bhopal, India).
- The polyvinyl chloride (PVC) pipe used in the EAHE is of uniform cross-section.
- The thickness of the pipe used in the EAHE is very small; hence, thermal resistance of pipe material is negligible.
- The temperature on the surface of the pipe is uniform in the axial direction.

1.3 Boundary conditions

The following boundary conditions were used in the one-dimensional model of the EAHE system.

1.3.1 Inlet boundary conditions

At the inlet of the EAHE pipe, the values of air flow velocity, v_a (m/s), and static temperature of air, T_{in} (°C), at inlet were to be defined. The thermodynamic properties (density and specific heat capacity) and transport properties (dynamic viscosity and thermal conductivity) of air were to be defined at static temperature of air at inlet.

1.3.2 Outlet boundary conditions

In a subsonic flow regime, the relative pressure at the outlet of the EAHE pipe was defined as equal to zero atm.

1.4 Wall

The temperature on the surface of pipe (wall) was uniform in axial direction and was defined as equal to earth's undisturbed temperature at Bhopal city (25.2 °C). No slip condition with smooth wall was assumed at the inner surface of the pipe.

1.5 Mass flow rate of air

The mass flow rate of air is an important parameter, and it must be known by the designer so that the selection of size and number of pipes can be initiated. There is no unique value of size and number of pipes which can meet the EAHE performance. So, the designer has to consider the best combination of the EAHE performance and pumping power required to ensure mass flow rate of air. For a pipe of diameter, D; air density, ρ; air flow velocity, v_a; and number of parallel pipes, N_p, the mass flow rate of air through a pipe, m_a, is given by:

$$\dot{m} = \frac{\frac{\pi}{4} D^2 \rho v_a}{N_p} \tag{1}$$

For the designer, these parameters have to be determined in such a way that the boundary conditions and the heat exchanger performance are met.

1.6 Earth's undisturbed temperature

The earth's undisturbed temperature is an important parameter in designing an EAHE system. Assuming homogeneous soil of constant thermal diffusivity, the temperature at any depth z and time t can be estimated by the following expression (Labs 1989):

$$T_{z,t} = T_m - A_s exp\left[-z\left(\frac{\pi}{365\alpha_s}\right)^{\frac{1}{2}}\right] \cos\left\{\frac{2\pi}{365}\left[t - t_o - \frac{z}{2}\left(\frac{365}{\pi\alpha_s}\right)^{\frac{1}{2}}\right]\right\} \tag{2}$$

where $T_{z,t}$ is the ground temperature at time t (s) and depth z (m), T_m is the average soil surface temperature (°C), A_s is the amplitude of soil surface variation (°C), α_s is the soil thermal diffusivity (m^2/s; m^2/day), t is the time elapsed from beginning of the calendar year (day), and t_o is the phase constant of soil surface (s; days).

It is very difficult to calculate accurately the value of earth's undisturbed temperature because the soil parameters are often unknown. Additionally, it is defined for mean soil properties. Hence, earth's undisturbed temperature is a hypothetical value which can be assumed as equal to the annual average soil surface temperature of a particular locality. The soil surface temperature is assumed equal to the ambient air temperature. So, the earth's undisturbed temperature for Bhopal (Central India) is defined as 25.2 °C which is equal to the annual average temperature for the same (source: Department of Meteorology, Bhopal).

2. METHODS

If the dimensions of the EAHE system are known, calculation of the heat transfer rate can be done either by using the log mean temperature difference (LMTD) method or the ε–number of transfer units (NTU) method. In this paper the ε–NTU method is used. The outlet temperature of air was determined by using effectiveness of EAHE (ε) which is a function of number of transfer units (NTU).

2.1 Heat exchanger effectiveness and NTU

In the earth–air heat exchanger, the medium used for transportation of heat is air only. The heat is released or absorbed by the air flows through the pipe walls by convection and from pipe walls to the surrounding soil and vice versa by conduction. If the contact of the pipe wall with the earth is assumed to be perfect and the conductivity of the soil is taken to be very high compared to the surface resistance, then the wall temperature at the inside of the pipe can be assumed to be constant. The expression of NTU depends on different types of flow configurations of the EAHE system. In this paper, the relationship for an evaporator or condenser (with a constant temperature on one side, i.e., wall) was used.

The total heat transferred to the air when flowing through a buried pipe is given by:

$$Q_h = \dot{m} C_p (T_{out} - T_{in}) \quad (3)$$

where $\dot{m}$ is the mass flow rate of air (kg/s), C_p is the specific heat of air (J/kg-K), T_{out} is the temperature of air at outlet of EAHE pipe (°C), and T_{in} is the temperature of air at inlet of EAHE pipe (°C).

Due to convection between the wall and the air, the transferred heat can also be given by:

$$Q_h = hA\Delta T_{lm} \quad (4)$$

where h is the convective heat transfer coefficient (W/m^2-K) and A is the internal surface area of the pipe (m^2).

The logarithmic average temperature difference (ΔT_{lm}) is given by ($T_{EUT} = T_{wall}$):

$$\Delta T_{lm} = \frac{T_{in} - T_{out}}{\ln\left[\frac{(T_{in} - T_{wall})}{(T_{out} - T_{wall})}\right]} \quad (5)$$

The temperature of air at the outlet of the EAHE pipe can be obtained in an exponential form as a function of the wall temperature and inlet air temperature by eliminating Q_h from Eqs. (3) and (4).

$$T_{out} = T_{wall} + (T_{in} - T_{wall})e^{-\left(hA/\dot{m}C_p\right)} \tag{6}$$

If a pipe of infinite length ($A = \infty$) is used, the air will be heated or cooled to the wall temperature. The effectiveness (ε) of EAHE for winter heating application can thus be defined as:

$$\varepsilon = \frac{T_{out} - T_{in}}{T_{wall} - T_{in}} = 1 - e^{-\left(hA/\dot{m}C_p\right)} \tag{7}$$

The non-dimensional group is called the number of transfer units (NTU):

$$\text{NTU} = \frac{hA}{\dot{m}C_p} \tag{8}$$

which gives

$$\varepsilon = 1 - e^{-\text{NTU}} \tag{9}$$

The effectiveness of earth–air heat exchanger is determined by the dimensionless group NTU. The variation in earth–air heat exchanger effectiveness as a function of number of transfer units is shown in Fig. 2. It was observed that with increase in value of NTU, the effectiveness also increases but the curve rapidly flattens. The relative gain in effectiveness is very small after the value of NTU becomes more than 3. There are a number of ways to construct an earth–air heat exchanger to obtain a given NTU and thus a desired effectiveness. Similar results were observed by De Paepe and Janssens (2003).

The influence of the design parameters on NTU can be studied in terms of heat transfer and pressure drop. The NTU consists of three parameters, namely, convective heat transfer coefficient (h), internal surface area of pipe (A) and mass flow rate of air ($\dot{m}$) which can vary.

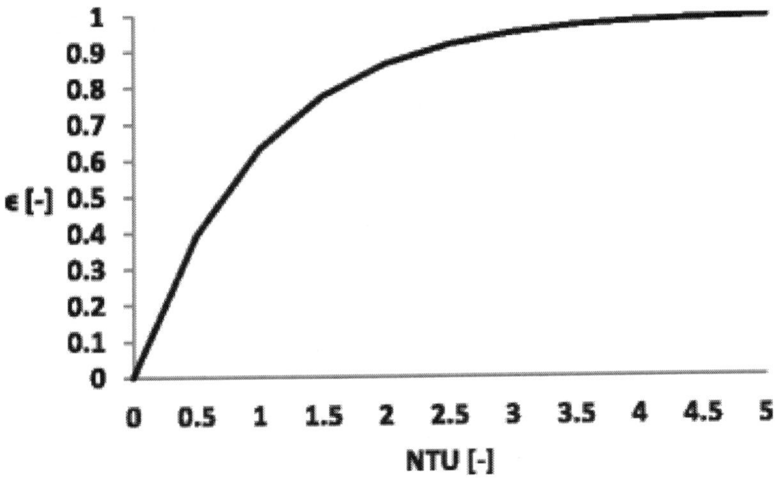

Figure 2. Earth–air heat exchanger effectiveness as a function of number of transfer units

The internal surface area of the pipe is a function of diameter, D, and length of EAHE pipe, L, both:

$$A = \pi DL \tag{10}$$

The convective heat transfer coefficient inside the pipe is defined by:

$$h = \frac{N_u K}{D} \tag{11}$$

where K is the thermal conductivity (W/m-K).

Zhang (2009) presented in his PhD thesis that in conventional earth-to-air heat exchanger (ETAHE) systems, it is typical to have buried ducts with 10 cm $< D_h <$ 40 cm and lengths longer than 20 m. Such sizes mean the ratios of the lengths to the hydraulic diameters (D_h) are at an order of magnitude of 100. The hydraulic diameter is defined as four times the ratio of the cross-section area to wetted perimeter of the cross-section.

$$D_h = \frac{4A}{P} \tag{12}$$

where A is the cross-section area and P is the wetted perimeter of the cross-section

The hydraulic diameter for a circular tube is simply the diameter of the tube. Therefore, it is reasonable to assume that the air flows are mostly fully developed in the EAHEs of such sizes and to adapt the corresponding empirical correlations to calculate the convection heat transfer coefficient (CHTC). In order to examine this assumption, eight Nusselt number (N_u) correlations used by other ETAHE simulation studies (Arzano and Goswami 1997; Bojic et al. 1997, cooling and heating; Singh 1994; De Paepe and Janssens 2003; Hollmuller 2003; Sodha et al. 1994; Benkert and Heidt 1997) were used. Since the correlations were all derived for fully developed turbulent air flow, ideally, they are expected to yield similar values for the same operating condition. The variation of Nusselt number with respect to Reynold's number for a typical design of conventional ETAHE was drawn using all eight correlations to calculate the CHTC, and very large differences were observed among the results of the eight correlations. This may be due to different experimental conditions, which were adopted to derive the correlations, for example, the surface roughness of the experimental ducts. The large discrepancies indicate that a suitable correlation has to be selected if one uses any of the existing models to simulate the performance of an EAHE system.

The EAHE system analysed in this paper consists of cylindrical pipes of 0.1016 m inner diameter made up of PVC with a total burial length of 19.228 m. Assuming that the internal surface of PVC pipes used in the EAHE system is smooth, the $N_u$correlations given by De and Janssens (2003) can be used to simulate the performance of the system.

$$N_u = \frac{f/8(R_e - 1000)P_r}{1 + 12.7\sqrt{(f/8)}(P_r{}^{2/3} - 1)} \tag{13}$$

(For turbulent flow in tubes with smooth internal surface)

where R_e is the Reynolds number, P_r is Prandtl number, and f is the friction factor for smooth pipes

$$\text{With } f = (1.82 \log R_e - 1.64)^{-2} \tag{14}$$

If $2300 \leq R_e < 5 \times 10^6$ and $0.5 < P_r < 10^6$

The Reynolds number is related to the average air velocity and diameter:

$$R_e = \frac{\rho v_a D}{\mu} \quad (15)$$

where v_a is the velocity of air through pipe (m/s), D is the diameter of the pipe (m), and μ is the dynamic viscosity of air (kg/m-s).

The Prandtl number is given by:

$$P_r = \frac{\mu c_p}{K} \quad (16)$$

where c_p is the specific heat of air (J/kg-K)

3. RESULTS AND DISCUSSION

The thermo-physical properties of materials used in design calculations of EAHE are shown in Table 1 (Bisoniya et al. 2014a; 2014b; 2014c; 2015).

Table 1. Thermo-physical properties of materials used in design calculations of EAHE

Material	Density (kg/m^3)	Specific heat capacity (J/kg-K)	Thermal conductivity (W/m-K)	Dynamic viscosity (kg/m-s)
Air at 16.7 °C	1.2185	1006	0.0253	1.804E-05
PVC	1380	900	0.16	-
Soil	2058	1843	0.54	-

The value of Reynolds number was calculated for air flow velocities of 2, 3.5, and 5 m/s and thermo-physical properties of air at 16.7 °C for winter heating application. The corresponding values of friction factor were calculated by using Eq. (14). The value of Prandtl number for thermo-physical properties of air at 16.7 °C was calculated as 0.717. After evaluation of Reynolds number, friction factor, and Prandtl number, the Nusselt number was calculated using Eq. (13) corresponding to air flow velocities of 2, 3.5, and 5 m/s.

The variation of Nusselt number with respect to Reynolds number is shown in Fig. 3. It was observed that the value of Nusselt number increases with increase in Reynolds number. Many other researchers (Nakamura and Tamotsu 2004; Luciu et al. 2009) also found that the value of Nusselt number increases with increase in Reynolds number.

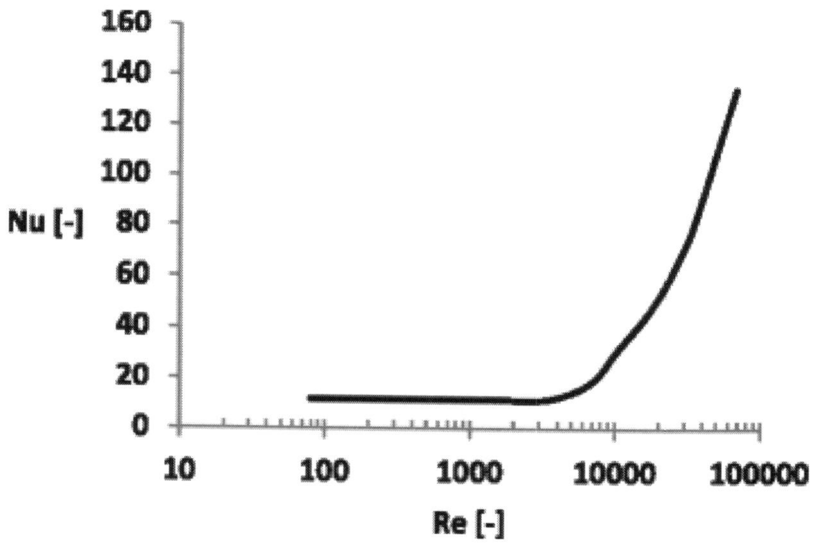

Figure 3. Variation of Nusselt number with respect to Reynolds number

The CHTC for different air flow velocities was calculated by using Eq. (11). The CHTC increases with increase in air flow velocity as shown in Fig. 4. Xiao et al. (2011) found that the variation in value of CHTC with a change in air flow velocity has shown similar trends.

The length, L, is an independent parameter influencing the NTU. There is a linear variation of NTU with length. Changing either the diameter, D, or mass flow rate, $\dot{m}$, changes the air velocity inside the tube. This result in a changing Reynolds number, R_e. D and $\dot{m}$ have thus no independent influence on the NTU. In general, lowering D raises the effectiveness; higher flow rates reduce the effectiveness. So, it is better to have several tubes of small diameter over which the flow rate is divided. Long tubes with a small diameter are profitable for the heat transfer. They, however, raise the pressure drop in the tubes, resulting in high fan energy.

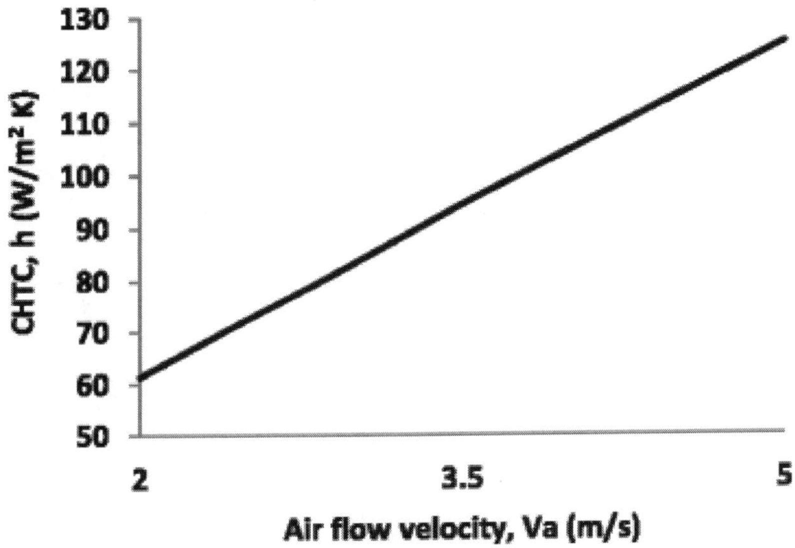

Figure 4. Convective heat transfer coefficient vs. air flow velocity

3.1 Influence on pressure drop

The pressure drop in a smooth pipe is given by:

$$\Delta p = f \frac{L}{D} \rho \frac{v_a^2}{2}$$

$$f = \frac{64}{R_e} \text{ If } R_e < 2300$$

$$f = (1.82 \log R_e - 1.64)^{-2} \text{ If } R_e \geq 2300 \quad (17)$$

It is noted from Eqs. (8) and (17) that both NTU and Δp are proportional to the length of the pipe, and the designer can use NTU/$L$and $\Delta p/L$ as the main performance measures to determine the required length of pipe for design purposes. The length of pipe, L, is an independent parameter which has a linear influence on pressure drop. The diameter of the pipe and air flow velocity has a combined effect on pressure drop. The decrease in air flow velocity and increase in diameter of pipe results in decrease in pressure drop. This is in disagreement with the thermal demand of a small diameter. In each case, a large number of pipes are beneficial. The combination of pipe length and diameter has to be optimized.

To evaluate the overall thermo-hydraulic performance of a specific configuration, the J-factor, introduced by De Paepe and Janssens (2003) was used. It is the ratio of the pressure drop to the NTU value and was shown to be a good performance metric of an earth–air heat exchanger.

$$J = \frac{\Delta P}{\mathrm{NTU}} \tag{18}$$

4. CONCLUSIONS

The earth–air heat exchanger is a promising technique which can effectively be used to preheat the air in winter and vice versa in summer. Many researchers have developed EAHE design equations and procedures. For a complete analysis of the EAHE system, the use of CFD is recommended but it is limited to those who have a good command over it. For the initial design of an EAHE system, the use of basic heat transfer equations is more suitable to determine the geometrical dimensions of the system. In this paper, the author has developed a one-dimensional model of the EAHE system. The method to calculate the EUT and more recently developed correlations for friction factor and Nusselt number are used to ensure higher accuracy in the calculation of heat transfer. The value of EUT for Bhopal (Central India) was calculated as 25.2 °C. It was observed that Nusselt number increases with increase in Reynolds number.

The design of earth–air heat exchanger mainly depends on the heating/cooling load requirement of a building to be conditioned. After calculation of heating/cooling load, the design of the earth–air heat exchanger only depends on the geometrical constraints and cost analysis. The diameter of pipe, pipe length, and number of pipes are the main parameters to be determined. With an increase in length of pipe, both pressure drop and thermal performance increase. A longer pipe of smaller diameter buried at a greater depth and having lower air flow velocity results in an increase in performance of the EAHE system.

COMPETING INTERESTS

The author declares that he has no competing interests.

REFERENCES

1. Arzano L, Goswami DY (1997) Performance Analysis of a Closed Loop Underground Air Tunnel for Residential Housing in a Hot and Humid Climate. Proceedings of the ISES Solar World Congress, Taejon, South Korea, August 1997

2. Badescu V, Isvoranu D (2011) Pneumatic and thermal design procedure and analysis of earth-to-air heat exchangers of registry type. Appl Energy 88:1266–1280

3. Benkert S, Heidt FD, Scholer D (1997) Calculation tool for Earth heat exchangers GAEA. Proceeding of Building Simulation, Fifth International IBPSA Conference, Prague, Czech Republic, pp 9–16, September 1997, 5

4. Bisoniya TS, Kumar A, Baredar P (2013) Experimental and analytical studies of earth–air heat exchanger (EAHE) systems in India: a review. Renew Sustain Energy Rev 19:238–246

5. Bisoniya TS, Kumar A, Baredar P (2014a) Cooling potential evaluation of earth–air heat exchanger system for summer season. Int J Eng Tech Res 2(4):309–316

6. Bisoniya TS, Kumar Anil, Baredar Prashant (2014b) Heating potential evaluation of earth–air heat exchanger system for winter season. Journal of Building Physics, 1–19.

7. Bisoniya TS, Kumar A, Baredar P (2014c) Parametric analysis of Earth-air heat exchanger system based on CFD modelling. Int J Power Renew Energy Syst 1:36–46

8. Bisoniya TS, Kumar A, Baredar P (2015) Energy metrics of earth–air heat exchanger system for hot and dry climatic conditions of India. Energy Build 86:214–221

9. Bojic M, Trifunovic N, Papadakis G, Kyritsis S (1997) Numerical simulation, technical and economic evaluation of air-to-earth heat exchanger coupled to a building. Energy 22(12):1151–115

10. De Paepe M, Janssens A (2003) Thermo-hydraulic design of earth-air heat exchangers. Energy Build 35:389–397

11. Gauthier C, Lacroix M, Bernier H (1997) Numerical simulation of soil heat exchanger storage system for greenhouse. Sol Energy 60(6):333–346

12. Hollmuller P (2003) Analytical characterisation of amplitude-dampening and phase-shifting in air/soil heat exchangers. Int J Heat Mass Transf 46:4303–4317

13. Hollmuller P, Lachal B (2001) Cooling and preheating with buried pipe systems: monitoring, simulation and economic aspects. Energy Build 33(5):509–518

14. Kanaris AG, Mouza AA, Paras SV (2006) Flow and heat transfer prediction in a corrugated plate heat exchanger using a CFD code. Chem Eng Technol 8:923–930

15. Labs K (1989) In: Cook J (ed) Passive cooling. Massachusetts, MIT Press, Cambridge

16. Luciu RS, Mateescu T, Cotorobai V, Mare T (2009) Nusselt number and convection heat transfer coefficient for a coaxial heat exchanger using Al_2O_3–water pH = 5 nanofluid. Bul Inst Polit Iasi 55:71–80

17. Mihalakakou G, Santamouris M, Asimakopoutos D (1994) Modeling the thermal performance of the earth to air heat exchanger. Sol Energy 53(3):301–305

18. Nakamura H, Tamotsu I (2004) Variation of Nusselt number with flow regimes behind a circular cylinder for Reynolds numbers from 70 to 30000. Int J Heat Mass Transf 47(23):5169–5173

19. Ozgener O, Ozgener L, Teste JW (2013) A practical approach to predict soil temperature variations for geothermal (ground) heat exchangers applications. Int J Heat Mass Transf 62:473–480

20. Sehli A, Hasni A, Tamali M (2012) The potential of earth-air heat exchangers for low energy cooling of buildings in South Algeria. Energy Procedia 18:496–506

21. Singh SP (1994) Optimization of earth-air tunnel system for space cooling. Energy Convers Manag 35(8):721–725

22. Sodha MS, Mahajan U, Sawhney RL (1994) Thermal performance of a parallel earth air-pipes system. Int J Energy Res 18(4):437–447

23. Su H, Liu X-B, Ji L, Mu J-Y (2012) A numerical model of a deeply buried air–earth–tunnel heat exchanger. Energy Build 48:233–239

24. T'Joen Christophe, Liu Liping, Paepe M De (2012) Comparison of earth-air and earth-water ground tube heat exchangers for residential application. International Refrigeration and Air Conditioning Conference, Paper 1209,

25. Wang Y, Dong Q, Liu M (2007) Characteristics of fluid flow and heat transfer in shell side of heat exchangers with longitudinal flow of shell side fluid with different supporting structures. International Conference on Power Engineering, Hangzhou, China, 23-27 October 2007

26. Xiao B, Wang G, Wang Q, Mohammed M, Sisson Jr Richard D, Yiming R (2011) An experimental study of heat transfer during forced air convection. J Mater Eng Perform 20(7):1264–1270

27. Zhang J (2009) Investigation of airflow and heat transfer in earth-to-air heat exchangers. Ph.D. thesis. Concordia University, Montreal, Quebec, Canada

CHAPTER 2

A New Adsorbent Composite Material Based on Metal Fiber Technology and Its Application in Adsorption Heat Exchangers

Ursula Wittstadt [1,*], Gerrit Füldner [1,†], Olaf Andersen [2,†], Ralph Herrmann [3,†] and Ferdinand Schmidt [4]

[1]Fraunhofer Institute for Solar Energy Systems ISE, Heidenhofstr. 2, D-79110 Freiburg, Germany

[2]Fraunhofer Institute for Manufacturing and Advanced Materials IFAM, Branch Lab Dresden, Winterbergstraße 28, D-01277 Dresden, Germany

[3]SorTech AG, Zscherbener Landstr. 17, D-06126 Halle/Saale, Germany

[4]Karlsruhe Institute of Technology (KIT), Institute of Fluid Machinery (FSM), Kaiserstr. 12, 76131 Karlsruhe, Germany

ABSTRACT

In order to achieve process intensification for adsorption chillers and heat pumps, a new composite material was developed based on sintered aluminum fibers from a melt-extraction process and a dense layer of silico-aluminophosphate (SAPO-34) on the fiber surfaces. The SAPO-34 layer was obtained through a partial support transformation (PST) process. Preparation of a composite sample is described and its characteristic pore size distribution and heat conductivity are presented. Water adsorption data obtained under conditions of a large pressure jump are given. In the next step, preparation of the composite was scaled up to larger samples which were fixed on a small adsorption heat exchanger. Adsorption measurements on this heat exchanger element that confirm the achieved process intensification are presented. The specific cooling power for the adsorption step per volume of composite is found to exceed 500 kW/m^3 under specified conditions.

KEYWORDS

adsorption heat exchanger (AdHEX); adsorption kinetics; SAPO-34; partial support transformation (PST); metal fibers; melt-extraction process

1. INTRODUCTION

In the development of adsorption heat pumps and chillers over the past decades, it has been a persistent goal to increase the volume-specific cooling or heating power to achieve more compact and light-weight units. The design of improved adsorption heat exchangers continues to take a central place in this effort.

Perhaps the simplest concept of an adsorption heat exchanger is a bed of adsorbent grains in a volume, part of which is occupied by some kind of heat exchanger. This configuration is widely used, and it can simply be obtained by pouring adsorbent grains into the volume in which the heat exchanger has been placed, and possibly densifying the bed by vibration. Whereas in standard applications of adsorbent beds (such as gas separation) the focus is on selectivity, for heat pumps and chillers the focus is on the heat of adsorption and the intensity of the heat transfer processes (during exothermal adsorption and endothermal desorption).

Therefore, the enhancement of heat conductivity within the adsorbent layer or bed is discussed by many authors (see e.g., references in [1]). The best heat transfer achievable as the limiting case of an adsorbent bed is obtained for a single layer of adsorbent grains on each heat exchanger surface, which is an approach taken, for example, in some zeolite/water heat pumps that have become available on the market recently ([2,3,4]). As an alternative to designs with a single adsorbent bed, additives have been proposed to enhance thermal conductivity. Wang *et al.* [5] have shown that a composite of polyaniline and adsorbents leads to a four-fold rise in thermal conductivity of the bed, whereas the compression of the bed improves it only by a factor of 1.5. Zhang [6] enhances the heat transfer in a cylindrical double tube by inserting fins in an adsorbent bed, whereas Demir *et al.* [7] and Rezk, Al-Dadah, Mahmoud, and Elsayed [1] investigate the effect of metal additives. Many experimental investigations on the performance of adsorbent beds in combination with finned tubes or lamella heat exchangers have been done [8,9,10,11]. Other authors propose to use commercially available plate-fin heat exchangers [12,13,14]. An overview on different concepts is given by Sharafian and Bahrami [15]. Here, the mass-specific cooling power ranges from 0.02 to 0.8 kW/kg of the adsorbent. Improvement of the thermal contact between the adsorbent and the heat exchanger, beyond the "packed bed" concept in its "single layer of grains" limit, can be achieved by increasing the contact surface through a binding agent or glue.

Coating the heat exchanger surface directly reduces the thermal contact resistance between the heat exchanger and the adsorbent. Jaeschke and Wolf [16] propose to glue a single layer of silica gel onto the surface of a finned-tube heat exchanger, whereas Freni *et al.* [17] report an increase by a factor of 10 to 20 in power density for a binder-based coating with the silica gel/calcium chloride composite material SWS-1L onto finned tubes compared to a granular bed. Other binder-based coatings show similar behavior [18,19].

Direct crystallization allows binding the pure adsorbent directly onto the heat exchanger surface. The absence of additional material augments the mass ratio of adsorbent to non-adsorbent material [20] and often shows a high stability [21,22]. In some cases, direct crystallization leads to a much more compact adsorbent layer (with a density close to the single-crystal density) than is achievable through a coating process. Different materials such as aluminum [23,24], copper [25], stainless steel [26], or graphite [27] can be used as a substrate.

Thick layers result in a higher adsorbent mass and thus an increase in Coefficients of Performance (COPs), but mass transfer may become the limiting factor. Especially when using water as a working fluid, the low operating pressure restricts the thickness of compact layers, whereas other working pairs like ammonia/activated carbon with a high operating pressure show less sensitivity to this restriction [28]. To overcome this issue, the surface area of the heat exchanger has to be enlarged considerably. The use of honeycomb structures [29], foams [25,30], or fiber material [31] has been proposed. Recently, Tatlier *et al.* [32] investigated coatings of zeolite A on stainless steel plates with different layer thickness (58–176 μm). They show that up to a certain layer thickness, the higher amount of adsorbents can compensate the reduction in rate of adsorption. Under the conditions investigated, an optimum was found for a layer thickness of 130–140 μm. An overview on recent developments in the field of coated adsorption heat exchangers and their characterization can be found in [33]. In this paper, we present a novel composite material of aluminum fibers coated by a silico-alumino-phosphate (SAPO-34) with a chabasite structure using a partial support transformation (PST) method as described by Bauer, Herrmann, Mittelbach, and Schwieger [23]. SAPO-34 has been shown to be a suitable adsorbent for heat transformation of low grade heat (<100 °C) [34,35,36,37]. Both the characteristics of the fiber material and the fiber-adsorbent composite are investigated. The application in adsorption processes is shown for small samples (30 mm × 30 mm × 3.9 mm). Finally, the feasibility of the scale-up to a new type of compact adsorption heat exchanger is presented and its water uptake behavior is examined.

2. PREPARATION OF COMPOSITE MATERIAL

2.1. Aluminum Fiber Structures

Short fibers of the aluminum alloy AlCu5 (aluminum with 5 wt% copper), specially designed for the required sintering process, have been produced by crucible melt extraction (CME) [38]. Porous bodies of various external shapes can be manufactured from such fibers through a process of fiber deposition and sintering, as described in detail in [39]. The alloy AlCu5 has been chosen for reasons of sinterability and good heat conductivity. Its density was found slightly higher than that of pure copper (3000 kg/m^3 instead of 2700 kg/m^3).

As liquid phase sintering has to be applied in order to achieve a good material bonding between the individual fibers, a sufficiently large super-solidus temperature interval is necessary in order to achieve a stable process window. The microstructure of the sintered fiber samples shows finely dispersed Al2Cu precipitates with a higher concentration at the grain boundaries. A patent on such super-solidus sintered aluminum alloy fiber structures made from eutectic alloys has been granted [40].

The CME process yields fibers with a low variation in length (mean lengths are determined by shape of extraction disc, typically 3 to 15 mm ±10%) and mean fiber diameters ranging from 50 to 250 μm. The geometric standard deviation of the fiber diameter is typically in the order of 2. Sintered fiber structures show a pore size distribution around a mean value that depends in a complex way on the geometric properties of the individual fibers and on the parameters of the subsequent processing [41]. In general, individual pore sizes may reach from 5 to 500 μm, with the mean pore size ranging from 50 to 250 μm, depending on the chosen porosity and the fiber diameter, *i.e.*, at the same porosity a finer diameter results in a finer mean pore size.

Similarly, the specific surface area may be adjusted by choosing a certain combination of porosity and mean fiber diameter. Attainable porosities range from 50% to 90%. The volume-specific surface area of these exclusively open-porous structures thus covers values from 2500 to 50,000 m^2/m^3.

The heat conductivity of such sintered metal fiber structures depends largely on the porosity level of the structure. Due to the manufacturing process, their heat conductivity has been found to be highly anisotropic [42]. Heat conduction along the direction of preferred fiber orientation may be two-to-three times higher than along the other directions. A factor influencing the degree of anisotropy is the fiber length, with longer fibers leading to stronger anisotropy. However, quantitative data on this hypothesis is not yet available.

The higher the relative density of the structure, the higher is the heat conductivity and the provided surface area for the zeolite coating. However, the accessibility of the structure for coating and transport of water vapor will be hindered when the resulting pore sizes are too small, therefore a good compromise was found with a porosity of 70% to 80% in conjunction with a fiber diameter of 100 to 200 μm, *i.e.*, for the AlCu5 fiber batch D108 with a mean equivalent diameter of 125 μm and a measured volume-specific surface area of 30.000 m^2/m^3, the resulting specific surface area amounts to 6000 and 9000 m^2/m^3 for a porosity of 80% and 70%, respectively. The measurement of the specific surface area is based on metallographic cross sections of a larger number of single fibers (typically >500). Computed tomography has shown that the specific surface values after sintering are very close to the metallographically measured values, meaning that the loss of surface due to sintering is typically in the range of 5% to 10%. Effective heat conductivities in this porosity range lie between 5 and 25 W/(m K), depending on the actual porosity of the sample and the direction of the heat flux in relation to the fiber orientation.

2.2. In Situ Crystallization of SAPO-34 on Aluminum Fiber Structures

In situ crystallization of SAPO-34 on aluminum supports has been reported by Bauer, Herrmann, Mittelbach, and Schwieger [23] and in US 8,053,032 B2 [24]. This process can also be described as a consumptive crystallization, since the Al support is partially transformed into the micro-porous adsorbent material SAPO-34. During the first part of this formation of a zeolitic aluminum phosphate, the metallic aluminum is dissolved to be available as the required Al source. Since the dissolution is controlled by the reaction conditions, it is possible to adjust the loss of fiber thickness to an amount that keeps the fibers itself and the fiber-fiber sinter connections intact. The solder for soldering the fiber package onto the heat exchanger was chosen to withstand the dissolution. On the other hand, the controlled crystal growth on the surface locally protects the Al, which leads to an advantageous micro-structured Al-adsorbent interface with an enhanced area for heat transport. This property also results in a very good mechanical bond between metal and adsorbent, which is especially important in sorption applications with cyclic temperature changes.

The SAPO-34 loading was adjusted to cover the fiber package with a mean thickness of approximately 45 µm. This amount of adsorbent keeps the porosity through the composite for water vapor transport while a high mass of adsorbent is realized (mean macropore diameter: 150 µm; see Table 1).

Table 1. Material properties of composite material.

Material propery	Unit	Value
Heat Conductivity	W/m K	9
Mean macropore diameter	m	150×10^{-6}
Macro porosity fibers	%	75
Macro porosity composite	%	29
Tortuosity	-	2–3
Permeability	m^2	1×10^{-10}

The structure of the zeolite coating was confirmed to be CHA by the XRD pattern (X'Pert Pro MPD, see Figure 1). The mass of the formed adsorbent was determined by measuring the mass loss during calcination at 520 °C and comparison with powder data (SAPO-34 powder: 18.7% calcination mass loss) as well as by water adsorption on the basis of the SAPO-34 isotherm (see Equation (2)). ICP-OES (Perkin Elmer Plasma 400) measurements of zeolite removed from the support show a composition of $(Si_{0.06}Al_{0.49}P_{0.45})\ O_2$. The density of the adsorbent layer amounts to around 1500 kg/m^3. This value was calculated from the adsorbent mass and the layer thickness (measured by cross section SEM, Hitachi S 2400).

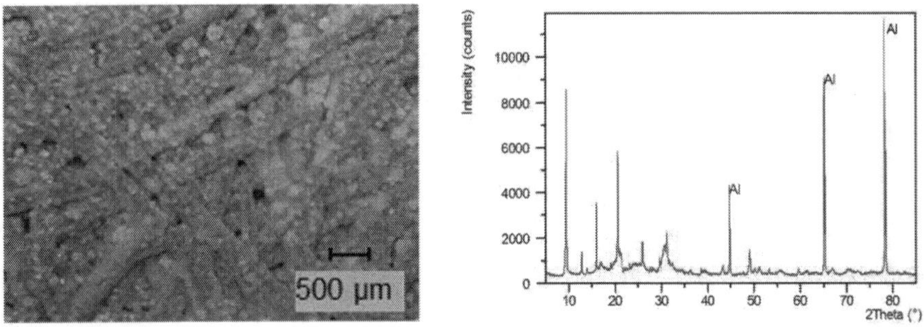

(a) (b)

Figure 1. (**a**) Microscopic view of a fiber-SAPO-34 sample; (**b**) XRD pattern of the SAPO-34 coated fiber sample; the reflexes of the fiber support are marked with Al.

2.3. Material Properties

The most important properties of the composite material and its application for adsorption heat exchangers (AdHEX) have been determined (see Table 1).

The thermal diffusivity of the composite material across the fiber structure has been determined with a flash method in a LFA 447 NanoFlash by Netzsch. With a specific heat capacity of 900 J/(kg K) [43] and a composite density of 1510 kg/m^3, the heat conductivity is found to be 9 ± 1 W/(m K) [44]. This is more than an order of magnitude higher than the heat conductivity of the adsorbent material and thus leads to a better heat transfer characteristic.

The pore size distribution of both the uncoated fibers as well as the composite material has been determined by mercury porosity under a protective gas atmosphere to prevent the formation of amalgam. A maximum of the pore size distribution is found at a pore width of around 330 μm for the pure aluminum fiber matrix and 150 μm for the macropores of the composite structure.

As viscous flow is the main transport mechanism (Knudsen number <<1) in the macropores of the composite material, the permeability κ can be determined by:

$$\kappa = \psi \cdot \frac{d_{pore}^2}{32 \cdot \mu} \quad (1)$$

with Ψ being the macro-porosity of the composite, μ the tortuosity, and d_{pore} the diameter of the macropores. By measurements in the slip flow regime (transition between Knudsen diffusion and viscous flow [45]), a tortuosity μ of 2 to 3 has been determined [44]. Thus, a value in the order of 1×10^{-10} m^2 can be calculated for the permeability.

3. ADSORPTION CHARACTERISTICS OF SMALL SAMPLE

The adsorption behavior of the small sample of the composite material (see Figure 2) has been investigated by a Large Pressure Jump (LPJ) method in a set-up described by Schnabel, Tatlier, Schmidt, and Erdem-Şenatalar [25]. The sample is placed on a cold plate in a vacuum chamber: thermostats control the temperature of the cold plate and the sample is desorbed (here: at 95 °C), either against a vacuum pump or against a temperature-controlled water flask, setting a certain condenser pressure for the desorption process and thus enabling the preparation of the sample at a well-defined initial adsorbate loading. The pressure in the vapor chamber is the saturation pressure of the water flask (here: 10 °C, *i.e.*, 12 mbar). The sample is cooled down to an adsorption temperature of 38.7 °C and adsorption is started by opening the valve between the vapor chamber and the measurement cell. The decrease in pressure in the combined volume over time is detected by a pressure sensor and the amount of water adsorbed by the sample can thus be determined with an accuracy of ±5% under the assumption that water vapor is an ideal gas [46] (see Figure 3 and Figure 4).

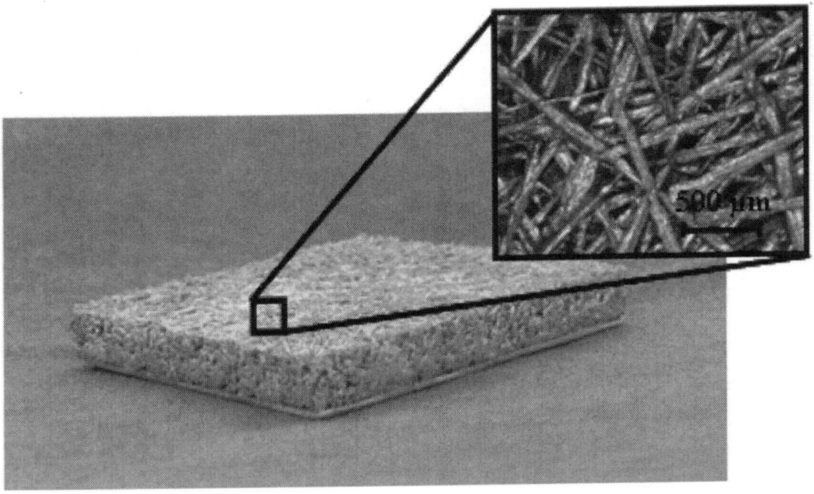

Figure 2. Sample of fiber-SAPO-34 composite material (0.03 m × 0.03 m × 0.0039 m), carrier: Al sheet of 0.4 mm thickness.

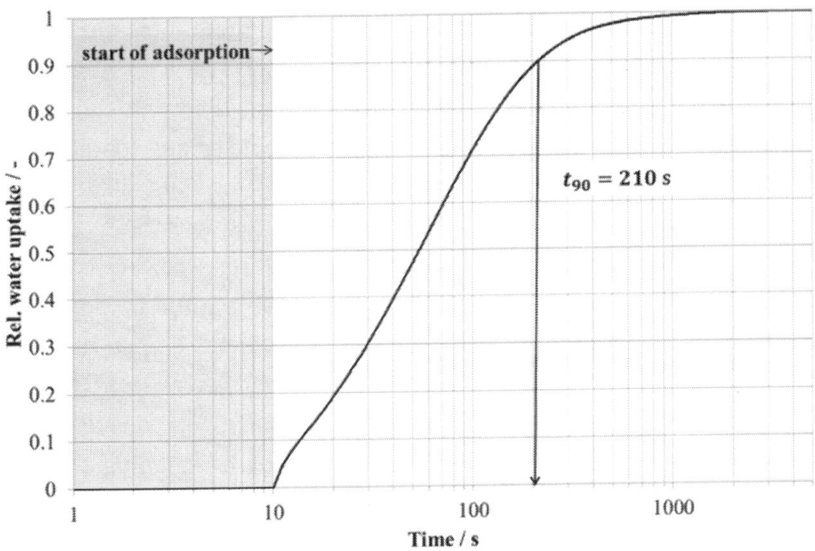

Figure 3. Water uptake (relative to final uptake) over time at 38.7 °C for small sample at a start loading of 0.06 g_{H2O}/g_{Ade}.

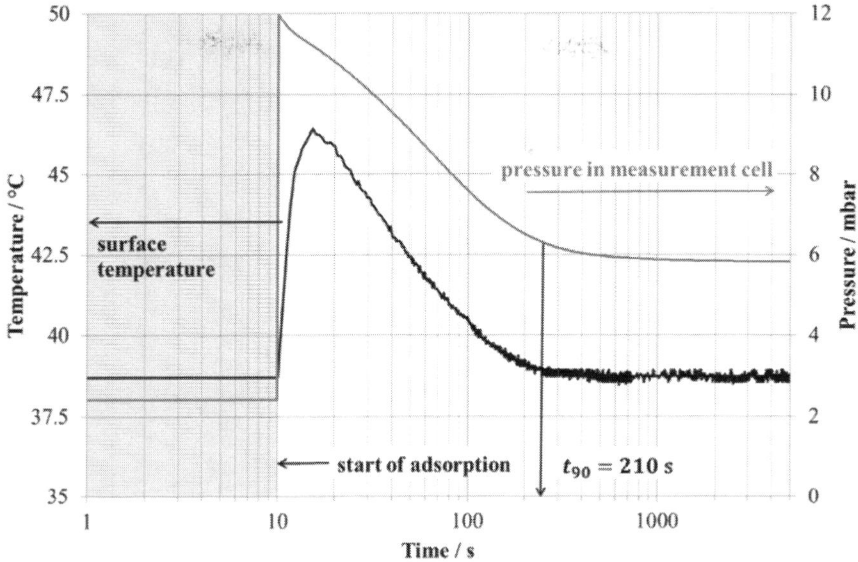

Figure 4. Surface temperature of sample and vapor pressure in measurement cell during water uptake as shown in Figure 3.

Figure 3 shows the water uptake over time measured at 38.7 °C. Further results are given in Table 2.

Table 2. Main parameters of small sample and prototype.

Parameter	Unit	Small Sample	Prototype
Volume			
Total	m^3	0.0039×10^{-3}	3×10^{-3} [a]
Profiles/support	m^3	0.0004×10^{-3}	0.495×10^{-3}
Fiber	m^3	0.0035×10^{-3}	0.315×10^{-3}
Mass			
Mass adsorbent	kg	2.2×10^{-3}	0.240
Total mass [b]	kg	6.8×10^{-3}	1.350
Specific values			
Mass adsorbent per total mass	kg/kg	0.32	0.18
Mass adsorbent per volume composite	kg/m^3	620	760

Notes: [a] Including void volume; [b] Measured at ambient conditions.

The sample adsorbs 180 mg of water in a short time: the time to reach 90% of the final uptake (t_{90}) is 210 s, which leads to a volume-specific cooling power density of over 500 kW/m^3 of the composite material.

4. LAB-SCALE ADSORBENT HEAT EXCHANGER (ADHEX)

The application of the newly developed composite material is demonstrated in a lab-scale adsorption heat exchanger.

4.1. Construction and Coating

A first scale-up to a heat exchanger device has been realized. The design is based on aluminum extrusion profiles which are welded into a collector tube. The fibers are sintered before being applied onto the heat exchanger. In the second step, fiber stripes are soldered onto the extrusion profiles. The whole fiber heat exchanger is then coated with SAPO-34 by a partial support transformation technique as described in Section 2.2.

The adsorption heat exchanger (AdHEX) is depicted in Figure 5. Its main parameters are given in Table 2.

4.2. Characterization

The kinetics of adsorption have then been measured in the set-up depicted in Figure 6. Its core part is a vacuum chamber in which the adsorption heat exchanger is placed. The water uptake is measured using a scale (WZA 8200, Sartorius, 8200 g ± 0.01 g) while the adsorption heat exchanger is connected to the hydraulic loop of the system. An evaporator is placed below the chamber and provides the water vapor to be adsorbed. During desorption, the same device is used as a condenser. Both the evaporator/condenser unit and AdHEX are equipped with volume flow and temperature sensors to determine the energy balance. The influence of different process parameters on the measurement of water uptake, such as temperature, pressure, and volume flow, has been investigated by Wittstadt *et al.* [47]. A maximum error of ±2 g in the measurement of water uptake has been determined.

(a) (b)

Figure 5. (**a**) Adsorption heat exchanger (AdHEX); (**b**) Detailed view: some parts of the fiber composite material still show poor contact to the surface of the heat exchanger.

Other than the measurement on small samples (see Section 3), adsorption takes place under quasi-isobaric conditions. After complete desorption, the connection between the evaporator/condenser and AdHEX is closed and both devices are cooled down to adsorption conditions. Due to the water vapor, which remains in the free volume of the vacuum chamber, this step is not isosteric. Since the pressure inside the chamber is measured, the water loading of the sorbent material can be calculated using the equilibrium data of SAPO-34. The following standardized Dubinin approach [48] is used for their description.

$$W(A) = \frac{a + c \cdot A + e \cdot A^2 + g \cdot A^3 + i \cdot A^4 + k \cdot A^5}{1 + b \cdot A + d \cdot A^2 + f \cdot A^3 + h \cdot A^4 + j \cdot A^5} \qquad (2)$$

with W being the adsorbed volume in cm^3/g adsorbent and *A* the adsorption potential in J/g adsorbent. The parameters a–k can be found in Table 3.

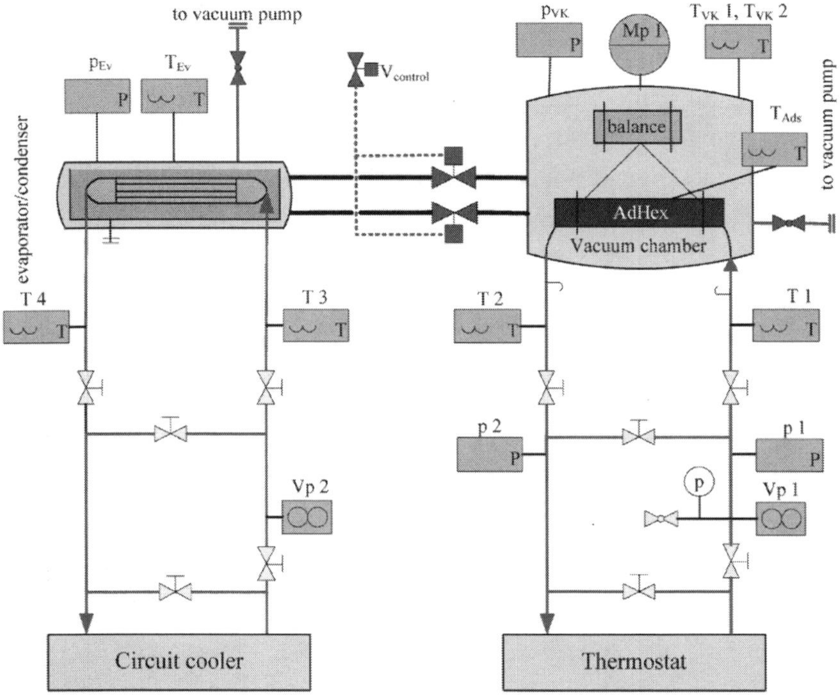

Figure 6. Schematic of the set-up for the characterization of small AdHEX.

Table 3. Parameters for the standardized Dubinin characteristic curve.

Parameter	Value	Parameter	Value
a	$0.33657\ cm^3/g$	g	$-1.26124 \times 10^{-8}\ (cm^3 \cdot g^2)/J^3$
b	$-0.00786\ g/J$	h	$3.86925 \times 10^{-11}\ g^4/J^4$
c	$-0.00270\ cm^3/J$	i	$9.22605 \times 10^{-12}\ (cm^3 \cdot g^3)/J^4$
d	$2.53508 \times 10^{-5}\ g^2/J^2$	j	$-8.10788 \times 10^{-15}\ g^5/J^5$
e	$8.39344 \times 10^{-6}\ (cm^3 \cdot g)/J^2$	k	$-2.48858 \times 10^{-15}\ (cm^3 \cdot g^4)/J^5$
f	$-4.35518 \times 10^{-8}\ g^3/J^3$		

A comparison of the water uptake behavior is given in Figure 7. The removal of the heat of adsorption influences the speed of the water uptake. With a very low volume flow of the heat transfer fluid (0.2 L/min, *i.e.*, residual time of 100 s) the 90% uptake time is 1520 s, around ten times higher than for the small sample. With a residence time of only

10 s (2 L/min) this time can be reduced to 630 s. A further intensification of the heat transfer to the fluid is not possible with the given heat exchanger geometry at a reasonable volume flow or pressure drop, respectively. As the flow channels have a small hydraulic diameter of around 3 mm, the flow stays laminar (Re = 12 and 121, respectively).

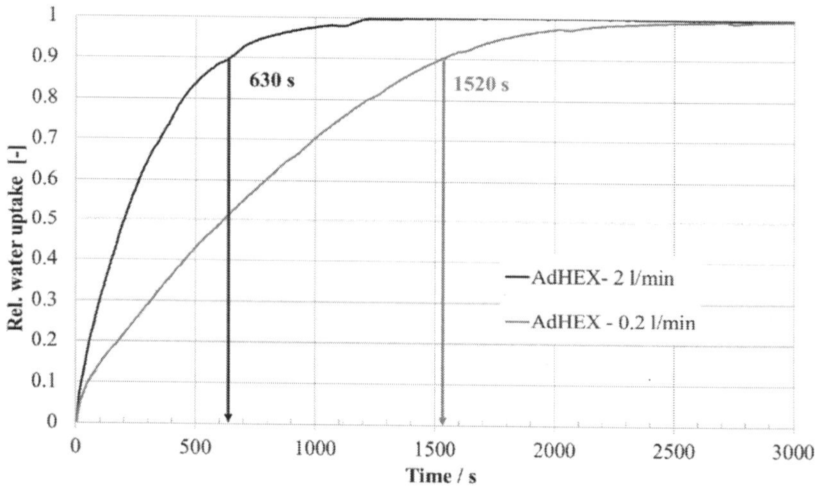

Figure 7. Comparison of water uptake behavior of the composite material in an AdHEX. The influence of the heat removal during adsorption in the AdHEX is shown for different volume flows of the heat transfer fluid.

The absolute water uptake at the end of adsorption differs within the expected error range. The main influencing parameters are uncertainties in the mass of active adsorbent and the correlation of the equilibrium data.

A direct comparison of the uptake behavior of the AdHEX to the one of the small sample is not possible as the measurements have been carried out under non-isobaric conditions and, hence, the uptake per g of adsorbent differs: the small sample adsorbs only around 40% of the water that is adsorbed per g of adsorbent in the AdHEX (0.11 g/g and 0.22 g/g, respectively), since the end pressure arrives at 5.8 mbar for the small sample, whereas for the AdHEX measurements, the end pressure lies at 11.9 mbar. Figure 8 shows the start and end loadings for the measurements on the 39 °C isotherm calculated from the equilibrium data of SAPO-34 (characteristic curve, see Equation (2)).

The relative water uptake behavior of the small sample can be used as a best-case scenario. Less heat has to be removed and the heat removal is close to ideal.

Although the uptake behavior is considerable slower for the AdHEX with t_{90} being three to seven times higher depending on the heat removal (see Figure 7), the volume-specific cooling power (VSCP) is slightly higher (see Figure 9). This is due to the fact that the composite material contains a higher amount of adsorbent material per volume. This outweighs the slower kinetics whereas the utilization of the sorbent material is lower in the AdHEX: the mass-specific cooling power (MSCP) is 0.77 kW/kg, which is 20% lower than the MSCP of the small sample (0.96 kW/kg). The rise-up time and sorption speed lie within the order of magnitude as reported in [19]. A direct comparison is difficult operation conditions as well as measurement procedure differences.

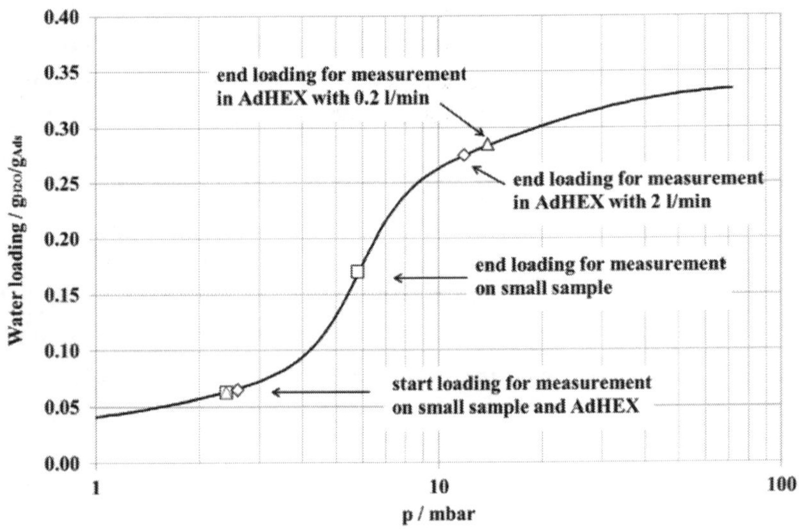

Figure 8. Isotherm at 39 °C for SAPO-34, calculated from the characteristic curve (Equation (2)). Open symbols indicate the start and end loading of all measurements. End loading of AdHEX differs slightly due to deviation in operation conditions (see Table 4).

Table 4. Parameters of the experimental characterization and main results of the water uptake behavior.

Parameter	Small Sample	AdHEX	
Volume flow of heat transfer fluid		0.2 L/min	2 L/min
Desorption temperature	95 °C	92 °C	95 °C
Adsorption temperature	38.7 °C	39.2 °C	39.8 °C
Starting pressure	2.4 mbar	2.4 mbar	2.6 mbar
End pressure	5.8 mbar	11.9 mbar	11.9 mbar
Pressure Jump	2.4–12.0 mbar	2.4–12.1 mbar	2.6–11.9 mbar
Condenser temperature/pressure	30 °C/42 mbar	24 °C/30 mbar	21 °C/25 mbar
Start loading (calculated)	0.06 g/g	0.06 g/g	0.06 g/g
End loading (calculated)	0.17 g/g	0.28 g/g	0.27 g/g
Rel. water uptake (calculated)	0.11 g/g	0.22 g/g	0.21 g/g
Abs. water uptake (calculated)	0.22 g	52.3 g	51.1 g
Rel. water uptake (measured)	0.09 g/g	0.23 g/g	0.22 g/g
Abs. water uptake (measured)	0.18 g ± 0.01 g	55 g ± 2 g	53 g ± 2 g
Time 90% uptake t_{90}	210 s	1520 s	630 s
Rise-up time t_{80}-t_{15}	118 s	1114 s	407 s
Sorption speed V_s	920 × 10^{-6} 1/s	187 × 10^{-6} 1/s	547 × 10^{-6} 1/s
Water uptake at t_{90}	0.16 g	49.9 g	47.4 g
Cooling power (t_{90})	2 W	80 W	184 W
Volume-specific cooling power (VSCP)			
per volume composite	545 kW/m^3	255 kW/m^3	585 kW/m^3
per volume composite + HEX	273 kW/m^3 (1)	98 kW/m^3	226 kW/m^3
per total volume AdHEX	55 kW/m^3 (2)	26 kW/m^3	59 kW/m^3
Mass-specific cooling power (MSCP)			
per mass adsorbent	0.96 kW/kg	0.33 kW/kg	0.77 kW/kg

(1): Fiber on profile as used for AdHEX (2): As (1), but plus same percentage of void volume as in AdHex.

In the AdHEX, the contact between the composite material and the heat exchanger surface for this first prototype is poor in some regions (see Figure 5b). A reduction of the thermal contact resistance might lead to a better performance if the mass transport into the adsorbent layer is not limiting the process. This question cannot be answered by just evaluating the experimental data, but has to be analyzed in a simulation with a detailed model of heat and mass transfer.

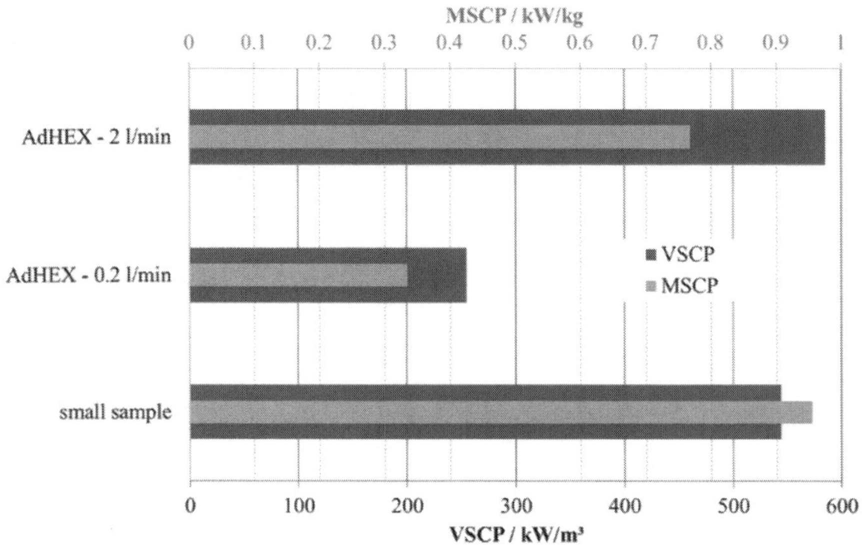

Figure 9. Mass-specific (MSCP) and volume-specific (VSCP) cooling power for the adsorption half-cycle calculated with 90% loading time of the composite material as a small sample and in an AdHEX.

5. CONCLUSIONS

A new composite material made from AlCu5 fibers which are coated with SAPO-34 by an *in situ* crystallization process has been developed. The high surface area of the fiber structures allows a high amount of adsorbent per volume (up to 760 kg/m^3) with a low adsorbent layer thickness. In combination with lowering the heat transfer resistance by soldering the fiber structure onto the support, a good heat transport characteristics i realized. A VSCP of 545 kW per m^3 of composite material measured on a small sample makes it favorable for adsorption chillers or heat pumps.

A scale-up from small samples to an AdHEX has been realized with an even higher amount of adsorbent mass and cooling power per volume of the composite. Further optimization is expected by enhancing the contact to the heat exchanger surface and by optimizing the heat transfer on the fluid side. Still, the vapor transport should be taken into account for the development of a more compact AdHEX design.

ACKNOWLEDGMENTS

Most of this work was carried out with funding of the Fraunhofer internal project "WISA Thoka" (Thermally driven high power density cooling processes). Funding for summarizing the contents and working out this paper from the German Federal Ministry of Education and Research (BMBF) in the framework of the project WasserMod (FKZ 03SF0469) is kindly acknowledged.

AUTHOR CONTRIBUTIONS

Olaf Andersen was responsible for the design, production, and characterization of the aluminum fiber structure. Ralph Hermann realized the direct crystallization of SAPO-34 on the fiber structure. Gerrit Füldner carried out the measurements of adsorption kinetics on the small samples. Ferdinand P. Schmidt led the project "WISA Thoka" and contributed to writing the paper. Ursula Wittstadt designed the Adsorption Heat Exchanger, carried out the experiments on it, and is the main author of the paper.

CONFLICTS OF INTEREST

The authors declare no conflict of interest.

REFERENCES

1. Rezk, A.; Al-Dadah, R.K.; Mahmoud, S.; Elsayed, A. Effects of contact resistance and metal additives in finned-tube adsorbent beds on the performance of silica gel/water adsorption chiller. *Appl. Therm. Eng.* **2013**, *53*, 278–284.

2. Wienen, J.; Neubert, M.; Lang, R.; Tiemeier, H. Results from field trial with gas heat pump Zeotherm by Vaillant. In *Thermally Driven Heat Pumps for Heating and Cooling*; Kühn, A., Ed.; Universitätsverlag der TU Berlin: Berlin, Geramny, 2013.

3. Núñez, T.; Mittelbach, W.; Henning, H.-M. Development of an adsorption chiller and heat pump for domestic heating and air-conditioning applications. *Appl. Therm. Eng.* **2007**, *27*, 2205–2212.

4. Schicktanz, M.; Núñez, T. Modelling of an adsorption chiller for dynamic system simulation. *Int. J. Refrig.* **2009**, *32*, 588–595.

5. Wang, L.; Zhu, D.; Tan, Y. Heat transfer enhancement of the adsorber of an adsorption heat pump. *Adsorption* **1999**, *5*, 279–286.

6. Zhang, L.Z. Design and testing of an automobile waste heat adsorption cooling system. *Appl. Therm. Eng.* **2000**, *20*, 103–114.

7. Demir, H.; Mobedi, M.; Ülkü, S. The use of metal piece additives to enhance heat transfer rate through an unconsolidated adsorbent bed. *Int. J. Refrig.* **2010**, *33*, 714–720.

8. Riffel, D.B.; Wittstadt, U.; Schmidt, F.P.; Nunez, T.; Belo, F.A.; Leite, A.P.F.; Ziegler, F. Transient modeling of an adsorber using finned-tube heat exchanger. *Int. J. Heat Mass Transf.* **2010**, *53*, 1473–1482.

9. Chang, K.S.; Chen, M.T.; Chung, T.W. Effects of the thickness and particle size of silica gel on the heat and mass transfer performance of a silica gel-coated bed for air-conditioning adsorption systems. *Appl. Therm. Eng.* **2005**, *25*, 2330–2340.

10. Chang, W.S.; Wang, C.C.; Shieh, C.C. Experimental study of a solid adsorption cooling system using flat-tube heat exchangers as adsorption bed. *Appl. Therm. Eng.* **2007**, *27*, 2195–2199.

11. Aristov, Y.I. Experimental and numerical study of adsorptive chiller dynamics: Loose grains configuration. *Appl. Therm. Eng.***2013**, *61*, 841–847.

12. Grisel, R.J.H.; Smeding, S.F.; de Boer, R. Waste heat driven silica gel/water adsorption cooling in trigeneration. *Appl. Therm. Eng.* **2010**, *30*, 1039–1046.

13. Freni, A.; Sapienza, A.; Glaznev, I.S.; Aristov, Y.I.; Restuccia, G. Experimental testing of a lab-scale adsorption chiller using a novel selective water sorbent "silica modified by calcium nitrate". *Int. J. Refrig.* **2012**, *35*, 518–524.

14. Santamaria, S.; Sapienza, A.; Frazzica, A.; Freni, A.; Girnik, I.S.; Aristov, Y.I. Water adsorption dynamics on representative pieces of real adsorbers for adsorptive chillers. *Appl. Energy* **2014**, *134*, 11–19.

15. Sharafian, A.; Bahrami, M. Assessment of adsorber bed designs in waste-heat driven adsorption cooling systems for vehicle air conditioning and refrigeration. *Renew. Sustain. Energy Rev.* **2014**, *30*, 440–451.

16. Jaeschke, S.; Wolf, M. Schichtverbund und seine Herstellung. German Patent DE102005038044, 10 August 2008.

17. Freni, A.; Russo, F.; Vasta, S.; Tokarev, M.; Aristov, Y.I.; Restuccia, G. An advanced solid sorption chiller using SWS-1L. *Appl. Therm. Eng.* **2007**, *27*, 2200–2204.

18. Munz, G.M.; Bongs, C.; Morgenstern, A.; Lehmann, S.; Kummer, H.; Henning, H.M.; Henninger, S.K. First results of a coated heat exchanger for the use in dehumidification and cooling processes. *Appl. Therm. Eng.* **2013**, *61*, 878–883.

19. Dawoud, B. Water vapor adsorption kinetics on small and full scale zeolite coated adsorbers: A comparison. *Appl. Therm. Eng.***2013**, *50*, 1645–1651.

20. Tatlıer, M.; Erdem-Şenatalar, A. The effects of thermal gradients in a solar adsorption heat pump utilizing the zeolite-water pair.*Appl. Therm. Eng.* **1999**, *19*, 1157–1172.

21. Bonaccorsi, L.; Calabrese, L.; Proverbio, E. Low temperature single-step synthesis of zeolite y coatings on aluminium substrates. *Microporous Mesoporous Mater.* **2011**, *144*, 40–45.

22. Freni, A.; Frazzica, A.; Dawoud, B.; Chmielewski, S.; Calabrese, L.; Bonaccorsi, L. Adsorbent coatings for heat pumping applications: Verification of hydrothermal and mechanical stabilites. *Innov. Mater. Process. Energy Syst.* **2010**, *50*, 1658–1663.

23. Bauer, J.; Herrmann, R.; Mittelbach, W.; Schwieger, W. Zeolite/aluminum composite adsorbents for application in adsorption heat pumps. *Int. J. Energy Res.* **2009**, *33*, 1233–1249.

24. Schwieger, W.; Thangaraj, S.; Scheffler, F.; Herrmann, R.; Reddy, M.; Mittelbach, W.; Bauer, J.; Schmidt, F.; Hennig, H.-M. Method for Production of a Substrate coated with a Zeolite Layer. US 8,053,032 B2, 29 October 2004.

25. Freni, A.; Bonaccorsi, L.; Proverbio, E.; Maggio, G.; Restuccia, G. Zeolite synthesised on copper foam for adsorption chillers: A mathematical model. *Microporous Mesoporous Mater.* **2009**, *120*, 402–409.

26. Schnabel, L.; Tatlier, M.; Schmidt, F.; Erdem-Şenatalar, A. Adsorption kinetics of zeolite coatings directly crystallized on metal supports for heat pump applications (adsorption kinetics of zeolite coatings). *Appl. Therm. Eng.* **2010**, *30*, 1409–1416.

27. Bonaccorsi, L.; Bruzzaniti, P.; Calabrese, L.; Freni, A.; Proverbio, E.; Restuccia, G. Synthesis of sapo-34 on graphite foams for adsorber heat exchangers. *Appl. Therm. Eng.* **2013**, *61*, 848–852.

28. Aristov, Y.I. Adsorption dynamics in adsorptive heat transformers: Review of new trends. *Heat Transf. Eng.* **2014**, *35*, 1014–1027.

29. Lang, R.; Westerfeld, T.; Gerlich, A.; Knoche, K.F. Enhancement of the heat and mass transfer in compact zeolite layers. *Adsorption* **1996**, *2*, 121–132.

30. Guilleminot, J.-J.; Choisier, A.; Chalfen, J.B.; Nicolas, S.; Reymoney, J.L. Heat transfer intensification in fixed bed adsorbers.*Heat Recover. Syst.* **1993**, *13*, 297–300.

31. Atakan, A.; Fueldner, G.; Munz, G.; Henninger, S.; Tatlier, M. Adsorption kinetics and isotherms of zeolite coatings directly crystallized on fibrous plates for heat pump applications. *Appl. Therm. Eng.* **2013**, *58*, 273–280.

32. Tatlier, M.; Munz, G.; Fueldner, G.; Henninger, S.K. Effect of zeolite a coating thickness on adsorption kinetics for heat pump applications. *Microporous Mesoporous Mater.* **2014**, *193*, 115–121.

33. Freni, A.; Dawoud, B.; Bonaccorsi, L.; Chmielewski, S.; Frazzica, A.; Calabrese, L.; Restuccia, G. *Characterization of Zeolite-based Coatings for Adsorption Heat Pumps*; Springer International Publishing: Cham, Switzerland, 2015. [

34. Henninger, S.K.; Schmidt, F.P.; Henning, H.-M. Water adsorption characteristics of novel materials for heat transformation applications. *Appl. Therm. Eng.* **2010**, *30*, 1692–1702.

35. Henninger, S.K.; Jeremias, F.; Kummer, H.; Schossig, P.; Henning, H.-M. Novel sorption materials for solar heating and cooling. *Energy Proced.* **2012**, *30*, 279–288.

36. Okamoto, K.; Teduka, M.; Nakano, T.; Kubokawa, S.; Kakiuchi, H. The development of AQSOA water vapor adsorbent and AQSOA coated heat exchanger. In Proceedings of the International Symposium on Innovative Materialsfor Processes in Energy Systems (IMPRESS 2010), Singapore, 29 November–1 December 2010; pp. 27–32.

37. Jänchen, J.; Ackermann, D.; Weiler, E.; Stach, H.; Brösicke, W. Calorimetric investigation on zeolites, $AlPO_4$'s and $CaCl_2$impregnated attapulgite for thermochemical storage of heat. *Thermochim. Acta* **2005**, *434*, 37–41.

38. Lehnert, F.; Lotze, G.; Stephani, G. Herstellung, eigenschaften und anwendung metallischer kurzfasern. *Materialwiss. Werkstofftech.* **1991**, *22*, 355–358.

39. Andersen, O.; Meinert, J.; Studnitzky, T.; Stephani, G.; Kieback, B. Highly heat conductive open-porous aluminium fibre based parts for advanced heat transfer applications. *Materialwiss. Werkstofftech.* **2012**, *43*, 328–333.

40. Andersen, O.; Studnitzky, T.; Kostmann, C. Bauteil Sowie Seine Verwendung. German Patent DE102007042494, 3 September 2007.

41. Andersen, O.; Stephani, G.; Meyer-Olbersleben, F.; Neumann, P. Properties of highly porous metal fibre components for high temperature applications. In Proceedings of the International Conference on Powder Metallurgy & Particulate Materials, Princeton, NJ, USA, 31 May–4 June 1998; Volume 3, pp. 87–97.

42. Veyhl, C.; Fiedler, T.; Andersen, O.; Meinert, J.; Bernthaler, T.; Belova, I.V.; Murch, G.E. On the thermal conductivity of sintered metallic fibre structures. *Int. J. Heat Mass Transf.* **2012**, *55*, 2440–2448.

43. Frazzica, A.; Füldner, G.; Sapienza, A.; Freni, A.; Schnabel, L. Experimental and theoretical analysis of the kinetic performance of an adsorbent coating composition for use in adsorption chillers and heat pumps. *Appl. Therm. Eng.* **2014**, *73*, 1020–1029.

44. Füldner, G. *Stofftransport und Adsorptionskinetik in porösen Adsorbenskompositen für Wärmetransformationsanwendungen*; Albert-Ludwigs-Universität Freiburg: Freiburg im Breisgau, Germany, 2015. (In German)

45. Kast, W. Diffusive mass transfer with superimposed frictional flow. *Int. J. Heat Mass Transf.* **2001**, *44*, 4717–4724.

46. Schnabel, L. Experimentelle und Numerische Untersuchung der Adsorptionskinetik von Wasser an Adsorbens-Metallverbundstrukturen. Ph.D. Thesis, Technische Universität Berlin, Berlin, Germany, 2009. (In German).

47. Wittstadt, U.; Jahnke, A.; Schnabel, L.; Sosnowski, M.; Schmidt, F.P.; Ziegler, F. Test facility for small-scale adsorbers. In Proceedings of the International Sorption Heat Pump Conference 2008, Seoul, Korea, 23–26 September 2008.

48. Núñez, T. *Charakterisierung und Bewertung von Adsorbentien für Wärmetransformationsanwendungen*; Albert-Ludwigs-Universität Freiburg: Freiburg im Breisgau, Germany, 2001. (In German)

CHAPTER 3

Heat Transport Study of the Laminar Heat Pipe Heat Exchanger

Wei-Keng Lin[1], Ke Chine Liaw[2], Min-Zen Tsai[2], Min-Gung Chu[2]

[1]Engineering & System Science Department, National Tsing Hua University, Hsinchu, Chinese Taipei;

[2]Tairone Energy Saving Technology Co., Ltd., Taoyuan, Chinese Taipei.

ABSTRACT

The purpose of this experiment is to analyze the performance of the Laminar Heat Pipe Heat Exchanger. The test samples were divided two groups, one is the metal corrugated sheet with heat pipe, and the other is the metal corrugated sheet only. By dipping these two group samples into hot water and to see the thermal image by Infrared thermal imager. The results shown the temperature risen of the sheet with heat pipe was faster than that of the sheet without heat pipe. In addition, the bigger of the temperature difference between hot water temperature and ambient temperature, the quicker of the temperature risen for the metal corrugated sheet, and the temperature of the metal corrugated sheet were homogenous as well. For example, when the water temperature is 30°C, ambient temperature is 20°C, the metal corrugated sheet with heat pipe rise rapidly to 26°C within 1 minute, while the metal corrugated sheet without heat pipe rise to 22°C only, this temperature difference would be more obvious with the increasing of the hot water temperature. When the hot water temperature is up to 40°C, the metal corrugated sheet with heat pipe rise rapidly to 31°C within 1 minute, while the metal corrugated sheet without het pipe is still rise up to 22°C only. When the hot water temperature is up to 50°C, the metal corrugated sheet with heat pipe rise rapidly to 33°C within 1 minute, while the sheet without heat pipe still keep at 22°C, the results shown the heat pipe affect the temperature rising speed is obvious, and it also implying heat pipe is a very important parameter for the heat transfer rate of the Laminar Heat Pipe Heat Exchanger.

KEYWORDS

Heat Pipe; Heat Exchanger; Metal Corrugated Sheet

1. INTRODUCTION

The heat exchanger is the indispensable equipment for mass and heat transport. In the petroleum, chemical, light industry, pharmaceuticals, energy, industrial production, the low temperature fluid need heat to high temperature fluid or conversely make the high temperature fluid turn to the low temperature fluid. All these processes would be achieved the energy transfer through liquid vaporization or vapor condensation. The medium of the heating or the condense process could be used gas to gas or liquid to gas. A variety of currently available heat exchanger classification can be divided into reversible (Recuperator), regenerative (Regenerator) and direct contact (Direct Contact Exchanger) [1]. The first type is Recuperator heat exchanger, **Figure 1**, heat is transfer from point B directly flow to point A through a thermal interface material. The common Recuperator heat exchanger include the air-cooled heat exchanger (**Figure 2**), shell and tube heat exchanger (**Figure 3**) [2] and the plate heat exchanger (**Figure 4**) [3,4]. Since the heat transfer for the first type Recuperator heat exchangermust pass through the interface, in order to improve the efficiency of the transmission, the materials used in heat exchangers is chosen the principle of material of high conductivity (κ) such as Cu (κ = 386 W/m·K) and Aluminum (κ = 204 W/m·K). The second type is the same as Recuperator heat exchanger, it through a working interface to transfer heat, but the transfer of heat does not occur at the same time, the heat transport mechanism shown as in **Figure 5**, the fluid A passing through the interface material within time (A). In the mean time, the heat is not releasing from the interface material until the fluid B passing through that interface material within time B, the energy is then complete transport from fluid A to fluid B, Rotary heat exchanger is the most famous example [5]. The third type of heat exchanger for direct contact heat exchanger, the transfer of heat by direct (or partial direct) contact with two working fluids. Cooling towers is a well-known example, even if the two working fluid is passed through direct contact with each other to achieve the energy transfer, but in the true operation, there is still need the base material and the flow guides to improve the mixing rate and mixing time of the fluid and thus result in the high heat transfer efficiency [6].

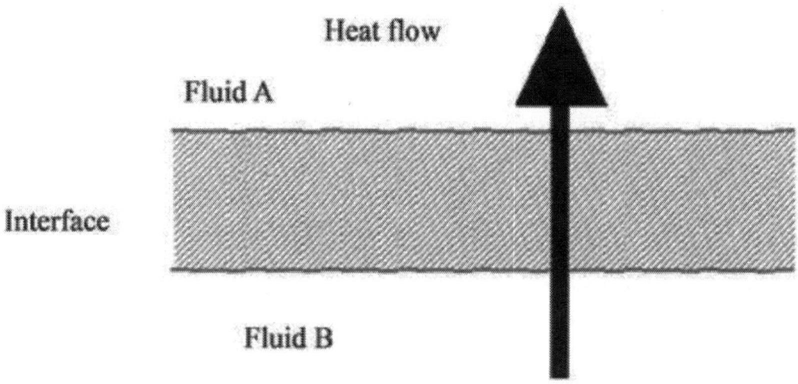

Figure 1. Operation principle of the recuperator heat exchanger.

Figure 2. Air-cooled heat exchanger.

Figure 3. Shell and tube heat exchanger.

Figure 4. Plate heat exchanger.

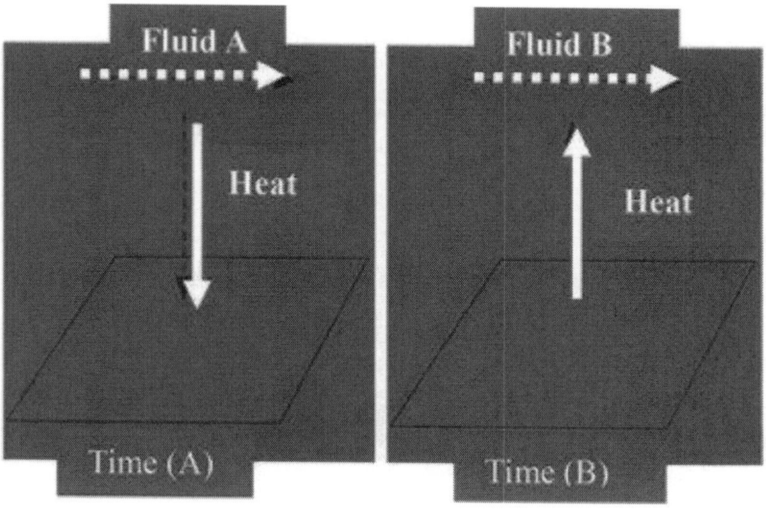

Figure 5. Operation principle of regenerative heat exchanger.

2. OPERATION THEORY

In this study, we are focus on the air cooling heat exchanger which is belonged to Recuperator heat exchanger type. The developed product in this study is to insert several heat pipes into the metal corrugated sheet namely “Laminar Heat Pipe Heat Exchanger”. The operation theory is as shown as in **Figure 6**, the gas in the indoor side is insulated from the gas in the outdoor side, and therefore, the gas flow is only taken away heat and not hybrid with the outdoor side gas and cause air pollution of the indoor air. The purpose of the heat pipe is to accelerate the heat conductance rate and achieve the temperature of the metal corrugated sheet homogeneous. The heat pipe is adhesive with the metal corrugated sheet as shown in **Figure 7**. In general, the heat conductivity of the heat pipe is 1000 times than that of the solid copper rod; therefore, the heat transport rate is 1000 time of the solid copper rod as well. It also implying when the heat pipe absorb heat at the evaporator section, the heat would be transported to the condenser section following the axial direction immediately. The high temperature vapor is released the heat in the condenser section and then turned out to be liquid form again and flow back to evaporator by the capillary force of the wick. When the heat pipe is transported heat from one side to another side periodically, the thermal energy is also transported to from heat pipe to metal corrugated sheet, therefore, the sheet obtained the heat very quickly as well.

3. EXPERIMENT AND PROCEDURE

The test two samples were shows as in **Figure 8**, the metal corrugated sheet is 400 × 400 mm, several 3 × 3 × 410 mm heat pipe were inserted into the metal corrugated sheet. The sample group which is no heat pipe namely “N-type”, while the group adhesive with heat pipe namely “W-type”. For the sake of the thermal image resolution, these two group metal corrugated sheets were coated with black color. To keep the temperature of the hot water reservoir constant, a 300 W heating electrical spoon with a constant temperature controller was used to heating the water shown in **Figure 9**. The Infrared thermal imager in this experiment was ThermoPro TP-8 shown as in **Figure 10**, the thermal imager was used to observed the 2D temperature distribution.

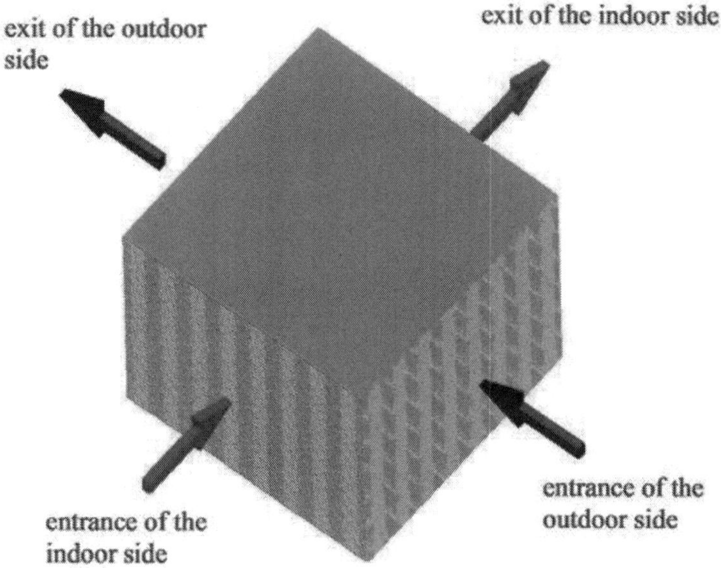

Figure 6. Operation principle of laminar heat pipe heat exchanger.

When the experiment start, add the water to the assigned level position, heating electrical spoon was then start to heat the water until the water temperature up to the required temperature which were 30°C, 40°C and 50°C. These two group metal corrugated sheets (W-type and N-type) were then dipped into the hot water reservoir with the same water level, **Figure 11**. In the meantime, another thermocouple was attached tightly on the surface of the other side of the metal corrugated sheet (cold side), recorded all the temperature distribution with respect to time.

4. RESULTS AND DISCUSSION

There were three kinds of the test model in this experiment, which were 30°C, 40°C and 50°C of the hot water reservoir temperature. The test samples were W-type and N-type, the ambient temperature keep at 20°C. The results were discussed respectively as following:

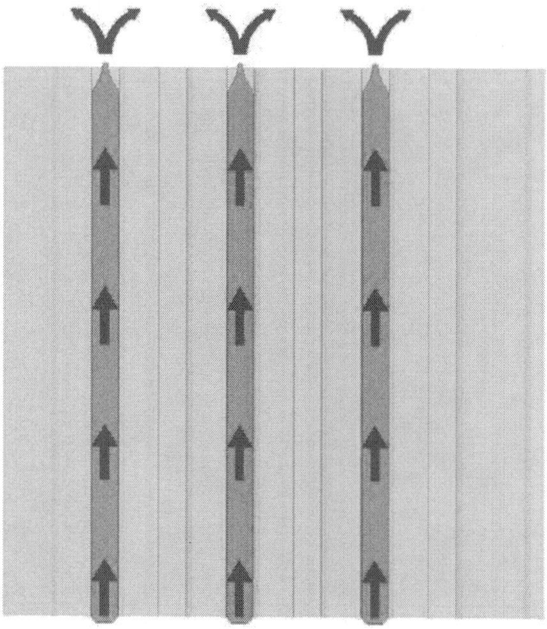

Figure 7. Orientation of the heat flow for the heat pipe in the corrugated sheet.

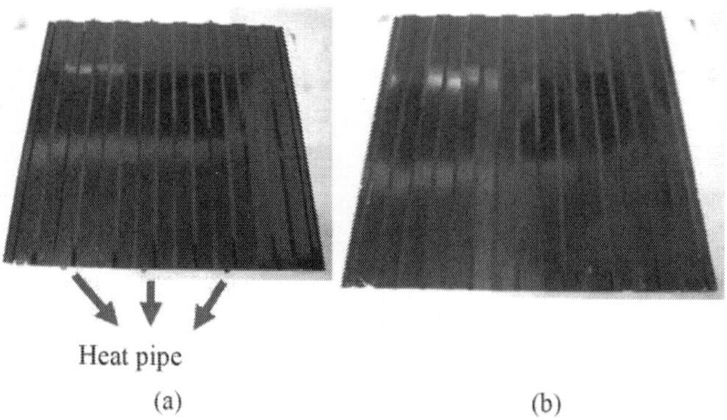

Figure 8. Metal corrugated sheet samples. (a) The metal corrugated sheet with Heat pipe (W-type); (b) The metal corrugated sheet without Heat pipe (N-type).

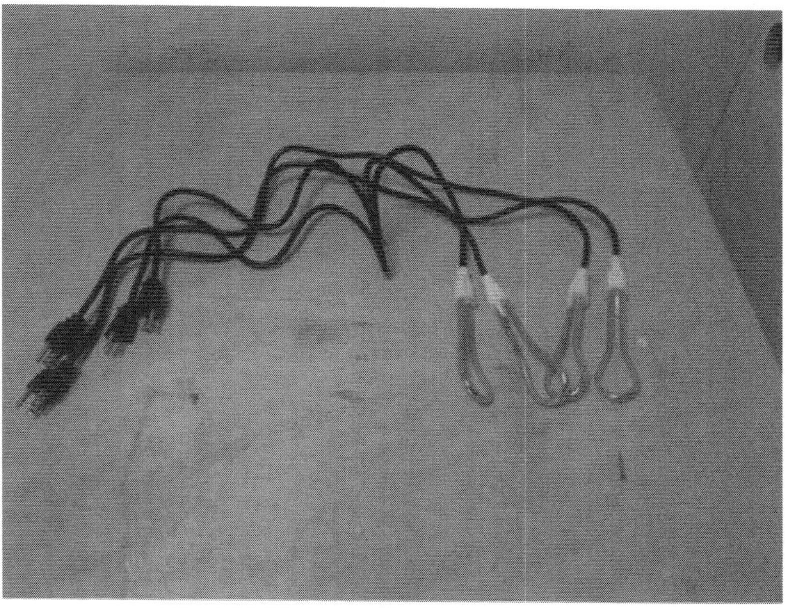

Figure 9. Heating electrical spoon.

Figure 10. Infrared thermal imager.

Figure 11. The metal corrugated sheet (W-type) with the hot water reservoir.

Model 1: ambient temperature: 20°C, water temperature: 30°C, test time period: 60 seconds.

As in **Figure 12**, the initial thermal image of the N-type at ambient temperature 20°C, water temperature 30°C. At time is zero, the cold side temperature (S1) of the metal corrugated sheet was 20.1°C which represented the thermal balance happened between metal sheet the ambient. After 60 seconds, **Figure 13**, the S1 temperature was up to 21.5°C and which was only 1.4°C difference compare with the initial temperature, it was also implying the temperature rising speed was very slow under the temperature difference 10°C between the water temperature and ambient temperature. The condition was completely changed when the metal corrugated sheet was inserted with heat pipe (W-type), **Figure 14** shown the same test condition as in the **Figure 13** but used the W-type instead. The cold side temperature (S1) of the metal corrugated sheet was 20.7°C at the initial state, however, after 60 seconds, the S1 temperature was up to 25.5°C, **Figure 15**, the temperature rising 4.8°C from the initial condition in this case. The result shown apparently that the 4.8˚C temperature difference for the W-type was faster than the temperature difference 1.4˚C for the Ntype. **Figure 16** show the comparison of the temperature rising between W-type and N-type at water 30˚C, the red line represented the temperature rising speed for the W-type, while the blue one represented the temperature rising speed for the N-type. The curve show the red one was faster than the blue one.

Model 2: ambient temperature: 20˚C, water temperature: 40˚C, test time period: 60 seconds.

As in **Figure 17**, the initial thermal image of the N-type at ambient temperature 20˚C, water temperature 40˚C. When at time is zero, the cold side temperature (S1) of the metal corrugated sheet was 23.6˚C. **Figure 18** showed the S1 temperature keep the same 23.6˚C after 60 seconds, there was almost no temperature change in this case. **Figure 19** showed the test condition for W-type at the initial condition, the cold side temperature (S1) of the metal corrugated sheet was 21.7˚C. But after 60 seconds, the S1 temperature was up to 30.6˚C, **Figure 20**, the temperature rising 8.9˚C from the initial condition.**Figure 21** show the comparison of the temperature rising between W-type and N-type at water 40˚C, the red line represented the temperature rising speed for the W-type, while the blue one represented the temperature rising speed for the N-type. The curve show the red one was faster than the blue one.

Model 3: ambient temperature: 20˚C, water temperature: 50˚C, test time period: 60 seconds.

As in **Figure 22**, the initial thermal image of the N-type at ambient temperature 20˚C, water temperature 50˚C. When at time is zero, the cold side temperature (S1) of the metal corrugated sheet was 20.2˚C, there was almost no temperature change in this case again. **Figure 23** showed the S1 temperature was up to 23.3˚C after 60 seconds and there is only 3.1˚C difference only.**Figure 24** showed the test condition for W-type at the initial condition, the cold side temperature (S1) of the metal corrugated sheet was 21.6˚C. But after 60 seconds, the S1 temperature was up to 32.8˚C, **Figure 25**, there was 11.2˚C temperature difference compare with the initial condition. The temperature rising speed 11.2˚C/min at water temperature 50˚C was also almost three times the speed 4.8˚C/min when water at 30˚C. **Figure 26** show the comparison of the temperature rising between W-type and N-type at water 50°C, the red line represented the temperature rising speed for the W-type, while the blue one represented the temperature rising speed for the N-type. Obviously, the curve show the red one was faster than the blue one.

Figure 12. Ambient temperature 20°C, water temperature 30°C, (N-type) initial thermal image.

5. CONCLUSIONS

1) The temperature rising speed for the metal corrugated sheet with heat pipe was faster than that of the metal corrugated sheet without heat pipe. The reason was that the heat pipe would accelerate the heat flow speed along the axial coordinate.

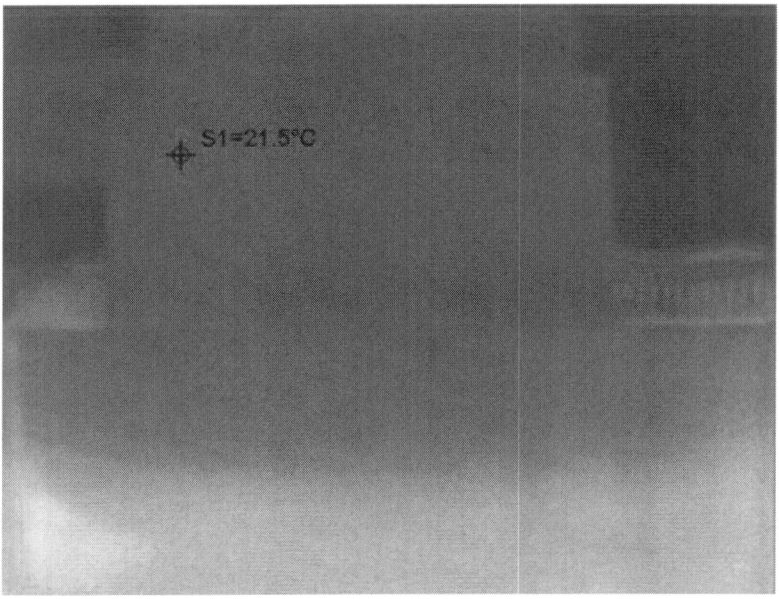

Figure 13. Ambient temperature 20˚C, water temperature 30˚C, (N-type) final thermal image after 60 seconds.

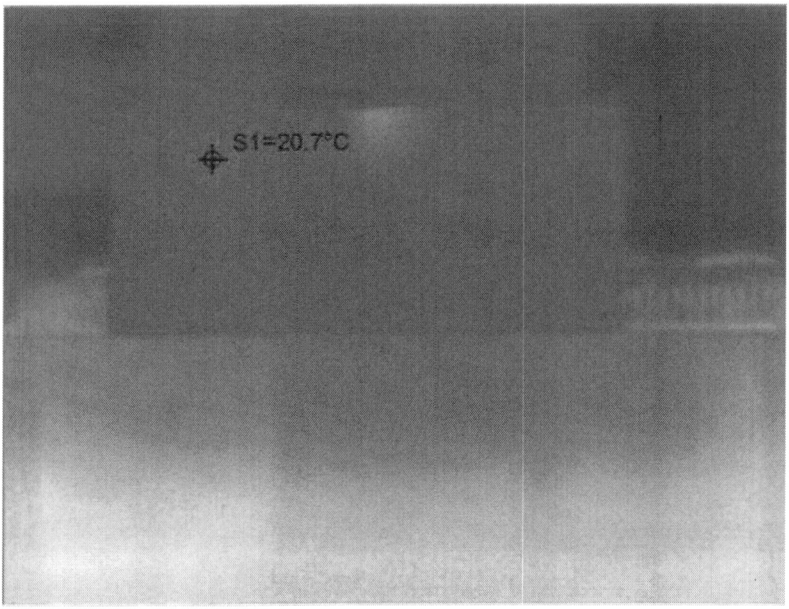

Figure 14. Ambient temperature 20˚C, water temperature 30˚C, (W-type) initial thermal image.

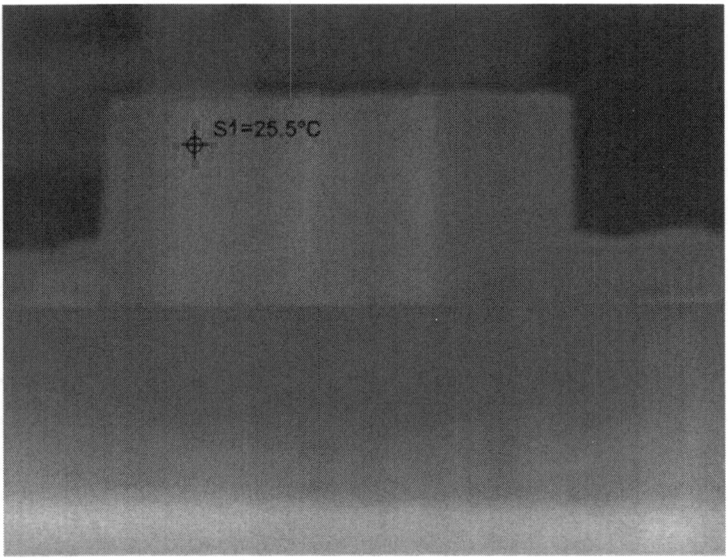

Figure 15. Ambient temperature 20˚C, water temperature 30˚C, (W-type) final thermal image after 60 seconds.

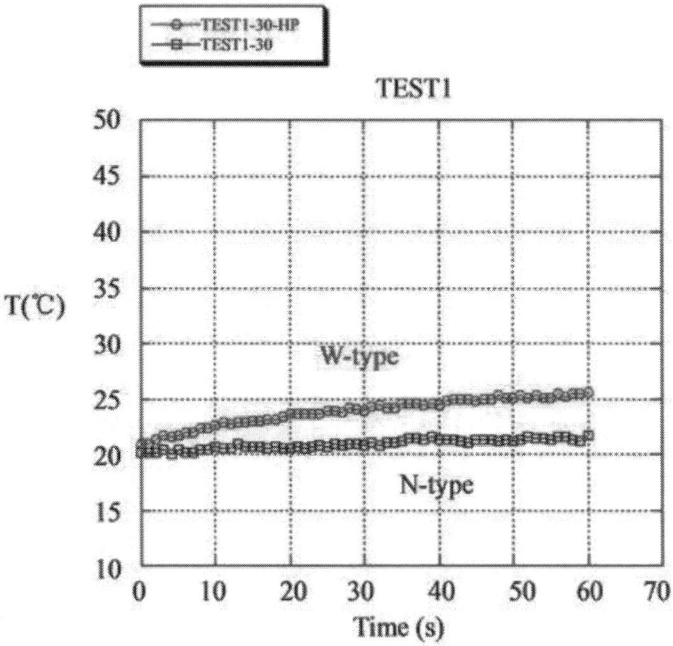

Figure 16. Comparison of the temperature rising between W-type and N-type at water 30˚C.

Figure 17. Ambient temperature 20˚C, water temperature 40˚C, (N-type) initial thermal image.

Figure 18. Ambient temperature 20˚C, water temperature 40˚C, (N-type) final thermal image after 60 seconds.

Figure 19. Ambient temperature: 20˚C, Water temperature: 40˚C, (W-type) initial thermal image.

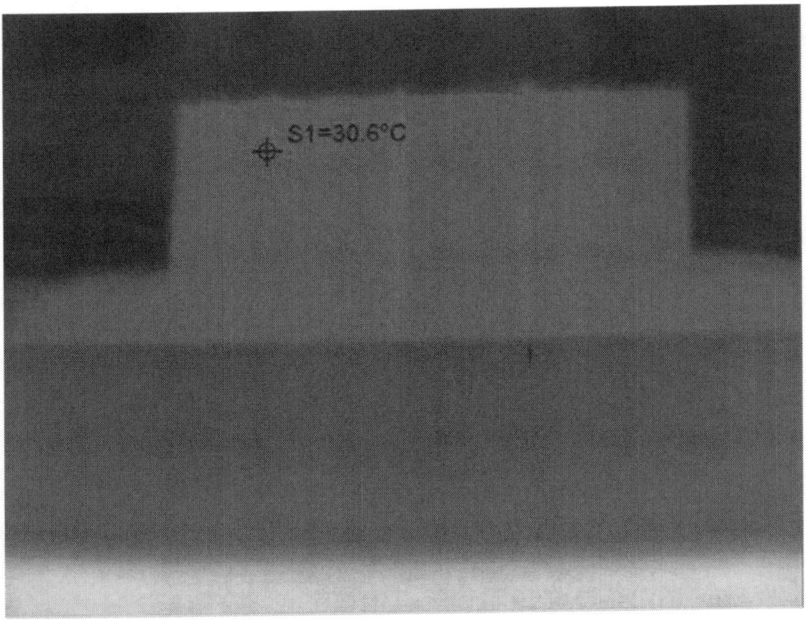

Figure 20. Ambient temperature: 20˚C, Water temperature: 40˚C, (W-type) final thermal image after 60 seconds.

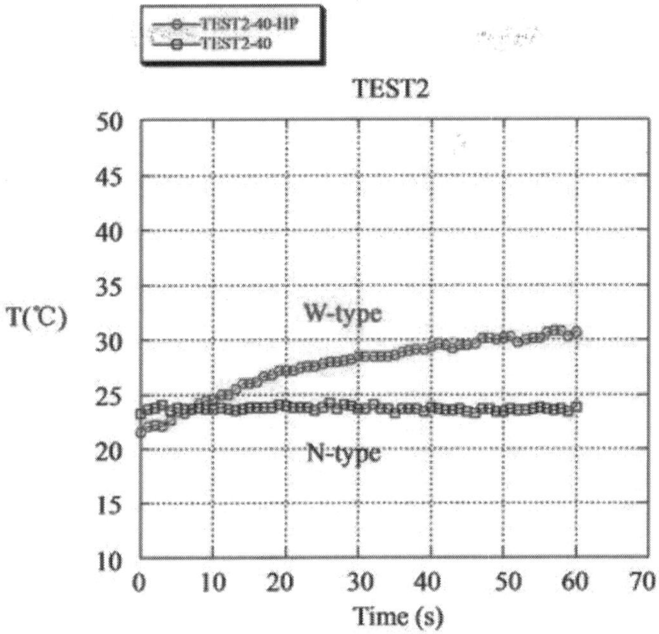

Figure 21. Comparison of the temperature rising between W-type and N-type at water 40°C.

2) When at constant ambient temperature, the bigger of the temperature different between the hot water temperature and ambient temperature, the faster of the temperature rising for the cold side temperature of the metal corrugated sheet, the more important of the heat pipe characteristic, **Figure 26** give the apparently example for the water temperature at 50°C.

3) For the case of the metal corrugated sheet with heat pipe, the sheet surface temperature would be more homogeneous shown as in Figures 15, 20 and 25. The reason why cause the surface temperature homogeneous was that the heat pipe not only transfer the heat following the axial direction of the sheet, but also spreading the heat to the X-Y direction of the sheet as well, thus result in the temperature of the surface of the sheet homogeneous.

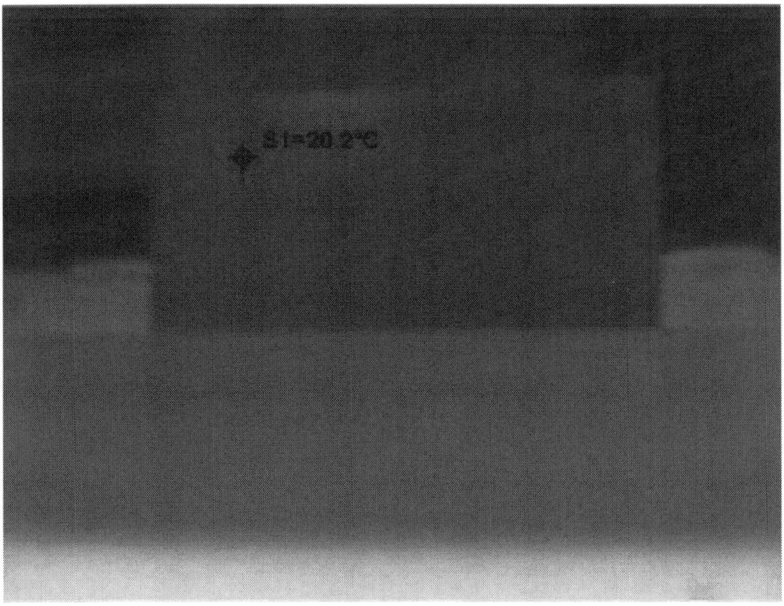

Figure 22. Ambient temperature: 20°C, Water temperature: 50°C, (N-type) initial thermal image.

Figure 23. Ambient temperature: 20°C, Water temperature: 50°C, (N-type) final thermal image after 60 seconds.

Figure 24. Ambient temperature: 20°C, Water temperature 50°C, (W-type) initial thermal image.

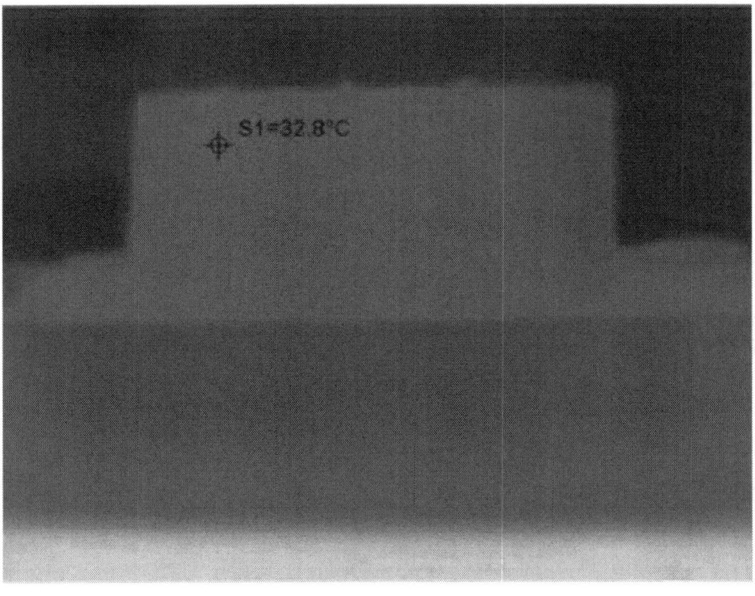

Figure 25. Ambient temperature: 20°C, Water temperature: 50°C, (W-type) final thermal image after 60 seconds.

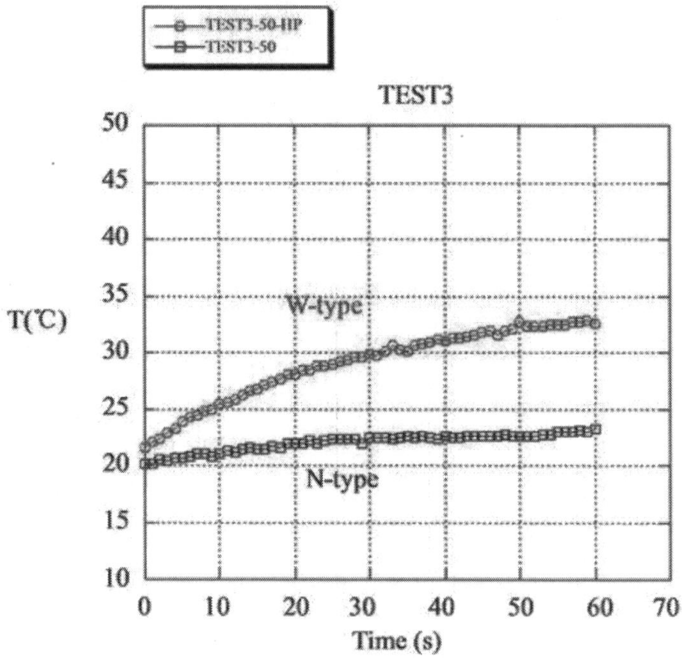

Figure 26. Comparison of the temperature rising between W-type and N-type at water 50°C.

ACKNOWLEDGEMENTS

This project (NSC-100-2622-E-007-006-CC2) is sponsor by Tairone Energy Saving Tech. Co., Ltd., The metal corrugated sheet provider, without their help, the experiment can't be carried out so smoothly completed.

REFERENCES

1. C. C. Wang, "Heat Exchanger Design," Wu-Nan Publication Co., Taipei, 2007.

2. D. B. Chang, "Heat Transfer Performance Enhancement of Shell and Tube Evaporator," Lectures in Nan-Tai University of Science and Technology, 2007.

3. J. P. Holman, "Heat Transfer," McGraw-Hill, New York, 1990.

4. G. F. Hewitt, "Executive Editor," Heat Exchanger Design Handbook, Begell House Inc., New York, 2002.

5. S. Kakac and H. Liu, "Heat Exchangers," CRC Press Ltd., London, 2002.

6. W. M. Rohsenow, J. P. Heartnett and Y. I. Cho, "Handbook of Heat Transfer," McGraw-Hill, New York, 1998.

CHAPTER 4

Designing coaxial ground heat exchangers with a thermally enhanced outer pipe

Jasmin Raymond[1]*, Serge Mercier[2] and Luc Nguyen[2]

[1]Institut national de la recherche scientifique, Centre Eau Terre Environnement, 490 de la Couronne, Québec (Québec) G1K 9A9, Canada.

[2]Versaprofiles, 130 rue Aubé, Saint-Lazare-de-Bellechasse, (Québec) G0R 3J0, Canada.

ABSTRACT

Background

Ground heat exchangers installed in boreholes are an expensive component of a ground-coupled heat pump system, where minimizing the borehole length with appropriate materials and configuration can reduce the overall cost of the system.

Methods

Design calculations performed analytically indicate that the coaxial pipe configuration can be more advantageous than the single U-pipe configuration to reduce the total borehole length of a system.

Results

A decrease of the borehole thermal resistance and an increase of the thermal mass of water contained in the coaxial exchanger helped to reduce borehole length by up to 23% for a synthetic building load profile dominated by cooling. The decrease of the borehole thermal resistance was achieved with an outer pipe made of thermally enhanced high-density polyethylene, where the thermal conductivity is 0.7 W m^{-1} K^{-1}.

Conclusions

The coaxial configuration requires further investigations of the technical barriers related to the installation of ground heat exchangers in the field.

KEYWORDS

Geothermal Heat pump, Ground heat exchanger, Borehole Coaxial Concentric Pipe, Thermally enhanced

1. BACKGROUND

Ground-coupled heat pump systems used for heating and cooling buildings are a highly efficient technology that takes advantage of the Earth's subsurface acting as a heat source or heat sink. The operation of heat pumps reduces the need for conventional energy. However, the systems are expensive because of the drilling or trenching required for installation of the ground heat exchangers (GHEs). Technological innovations can help reduce the length of GHEs for building energy needs to be fulfilled at lower installation cost. Reducing installation costs is particularly important for vertical systems with boreholes that tend to be more expensive than horizontal systems (Canadian GeoExchange Coalition 2010). Significant advances in addressing this market barrier can reduce the payback period of geothermal systems and increase the shares of the geothermal sector on the global heating and cooling market.

To determine the required length of vertical GHEs, one of the various parameters considered in the equation is the borehole thermal resistance (Bernier 2000), which is the ability of the GHE to resist heat transfer. Selecting appropriate materials and configurations to optimize the borehole thermal resistance can help decrease borehole length. To reduce the borehole thermal resistance of GHEs made with one or more U-pipe, advances were made to develop space clips holding pipes separately, thermally enhanced grout (Kavanaugh and Allan 1999; Allan and Kavanaugh 1999; Carlson 2000; Borinaga-Treviño et al. 2013a,2013b) and thermally enhanced pipe (Raymond et al. 2011a, 2011c). Further work has recently been conducted on coaxial GHEs, where the exchanger consists of two pipes imbricated into each other. The impact of the flow rate and the pipe diameter and thermal conductivity on the heat transfer rate was investigated with numerical simulations (Zanchini et al. 2010). Detailed experimental work was performed in the field to evaluate the borehole thermal resistance of different coaxial GHE configurations, where the external flow channel was made of several small pipes (Acuña et al. 2011) or a flexible liner molding the borehole wall and avoiding the use of backfill material (Acuña and Palm 2013). An analytical solution was also developed to estimate the vertical temperature distribution in the ascending and descending pipe of a coaxial GHE to improve the analysis of thermal response tests (Beier et al. 2013).

While previous research showed potential advantages of using coaxial GHEs, none of the above authors have investigated the possible bore length reduction associated with coaxial configurations. When compared to GHEs with a single or a double U-pipe that are commonly used, coaxial GHEs can contain a larger amount of water allowing greater heat storage to buffer the building peak loads. Thermally enhanced pipes made of carbon nanoparticles and high-density polyethylene (HDPE) can further reduce the borehole thermal resistance of coaxial GHEs. Design calculations were performed in this study using analytical models with the objective of evaluating the possible bore length reduction that can be achieved when using coaxial GHEs for ground-coupled heat pumps. The focus was given to pipe parameters to verify how this part of the heat exchanger can be improved. Calculations of borehole thermal resistance and heat storage capacity are initially presented to facilitate the design of coaxial GHEs. A synthetic building load profile was then used to size a ground-coupled heat pump system with either U-pipe or coaxial GHEs.

2. METHODS

2.1 GHE materials and configurations

In Northeast America, single U-pipe GHEs are commonly installed in boreholes with a diameter and length of 152.4 mm (6 in.) and 152.4 m (500 ft; Figure 1a). Occasionally, a double U-pipe (Figure 1b) can be installed in the boreholes, but this practice seems to be more popular in Europe. The borehole is filled with thermally enhanced grout made of water, silica sand, and bentonite that has a thermal conductivity approximately equal to 1.7 W m^{-1} K^{-1}. Again, this practice is quite common in America while this is not always the case in Europe, for example, in Sweden (Gustafsson and Westerlund 2011), where groundwater-filled boreholes are used. The pipe is made of HDPE and has a nominal diameter of 31.8 mm (1.25 in.). The standard dimension ratio (SDR) of the pipe is commonly 11, which is obtained by dividing the pipe outer diameter with the pipe thickness. The thermal conductivity of the HDPE forming the pipe is assumed to be 0.4 W m^{-1} K^{-1}. The HDPE can be mixed with carbon nanoparticles to make a thermally enhanced pipe with a thermal conductivity of 0.7 W m^{-1} K^{-1} (Raymond et al. 2011a). Versaprofiles is the first company to commercialize such a pipe, sold under the name Geoperformx. Space clips can be used to separate the pipes from each other, reducing thermal short circuiting and consequently decreasing the borehole thermal resistance. The heat carrier fluid made of water and antifreeze enters one side of the U-pipe and exits the other at a velocity sufficient to ensure turbulent flow and facilitate heat transfer.

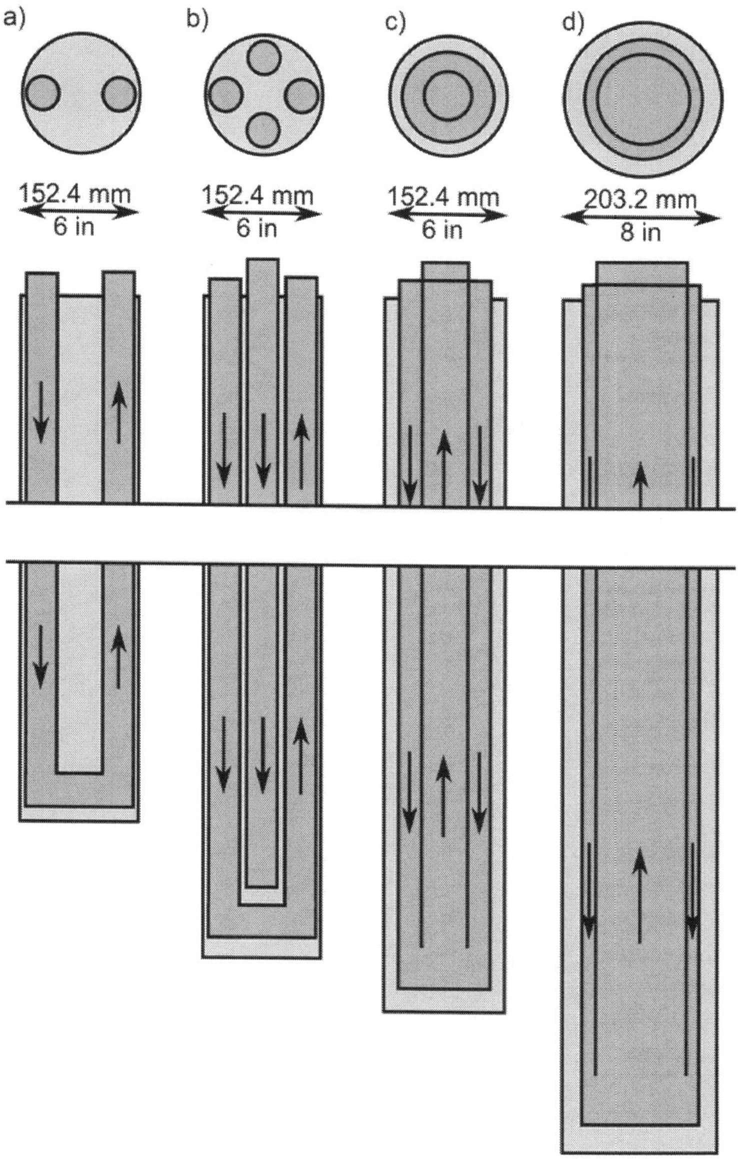

Figure 1. U-pipe (a,b) and coaxial (c,d) GHE configurations used for design calculations. The GHE length is not to scale.

The alternative coaxial design considered in this study involves the use of two pipes installed in each other to make the GHE (Figures 1c, d). Grout fills the space between the borehole wall and the outer pipe. The boreholes selected had a diameter equal to 152.4 and 203.2 mm (6 and 8 in.), respectively, with outer and inner pipes having a nominal diameter equal to 101.6 and 50.8 mm (4 and 2 in.) and 152.4 and 101.6 mm (6 and 4 in.). The heat carrier fluid enters the annulus and exits the inner pipe. Only pure water was selected for the heat carrier fluid to maintain a low pumping power at the higher flow rate required for turbulence to occur in the GHE annulus.

Borehole thermal resistance calculations were performed to evaluate the impact of the water flow rate, the borehole length, the backfilling material as well as the pipe thermal conductivity and SDR according to the range of parameters given in Table 1. Pure water properties were used for the heat carrier fluid. A subsurface thermal conductivity of 2.5 W m^{-1} K^{-1}, a volumetric heat capacity of 2.5 MJ m^{-3} K^{-1}, and an undisturbed temperature equal to 10°C were assumed for all simulations of the GHE temperature. Subsurface parameters were kept constant to verify how design parameters, especially those related to piping, can improve GHE performances.

Table 1. Range of each parameter used for borehole thermal resistance calculation

Parameter	Minimum	Maximum	Average
Water flow rate (L s^{-1})	0.5	6	3.25
Borehole length (m)	50	250	150
Backfilling thermal conductivity (W m^{-1} K^{-1})	0.6	2.8	1.7
Pipe thermal conductivity (W m^{-1} K^{-1})	0.1	0.7	0.4
Pipe outside diameter (mm)/thickness (mm)	SDR-11	SDR-17	SDR-13.5
31.75-mm (1.25 in.) nominal diameter	42.0/3.8	Not used for U-pipe	Not used for U-pipe
50.8-mm (2 in.) nominal diameter	59.7/5.5	60.3/3.6	59.8/4.5
101.6-mm (4 in.) nominal diameter	113.1/10.4	114.3/6.7	113.3/8.5
152.4-mm (6 in.) nominal diameter	166.4/15.3	168.3/9.9	166.8/12.5

2.2 Design calculations

The analytical approach used by the commercial programs Earth Energy Design (EED) and Ground Loop Heat Exchanger Professional (GLHEPro) were combined to properly calculate the borehole length required for a ground-coupled heat pump system with coaxial GHEs. Three-dimensional borehole thermal resistances of coaxial GHEs were initially calculated with EED according to Hellström's method (1991). The calculation includes a two-dimensional outer resistance R_b (m K W^{-1}) determined from the flow of water in the annulus and the thermal conductivity and dimensions of the outer pipe and the backfilling material. A two-dimensional internal resistance R_a (m K W^{-1}) is additionally determined from the flow of water in the inner pipe as well as the thermal conductivity and dimensions of the inner pipe. Both the outer and internal resistances are then used to determine an effective three-dimensional borehole thermal resistance R∗bRb∗ (m K W^{-1}):

$$R_b^* = R_b\left(1 + \frac{H^2}{3(\dot{m}_w c_w)^2 R_a R_b}\right) \tag{1}$$

where H (m) is the GHE length, $\dot{m}_w$ (kg s^{-1}) is the mass flow rate of water circulating inside the GHE, and c_w (J kg^{-1} K^{-1}) is the water specific heat capacity. Equation 1 assumes a constant heat injection rate along the borehole depth, which is convenient but departs from field observations (Beier et al. 2013). Using Hellström's method (1991) implemented in EED, it is possible to take into account the internal resistance, which can be significant for long boreholes typically drilled for coaxial GHEs. Calculating borehole thermal resistances with the program GLHEPro involves a two-dimensional approach that would not be suitable for determining the impact of increasing the borehole length or insulating the inner pipe.

Effective three-dimensional borehole thermal resistances of GHEs with a single and a double U-pipe were also calculated with Equation 1 and compared with the resistances of coaxial GHEs. In that case, the two-dimensional borehole thermal resistance was determined using the multipole method, with recent improvements about boundary conditions (Claesson and Hellström2011).

For sizing and simulation purposes, the average water temperature $\bar{T}_w$ (°C) inside the GHEs was calculated with GLHEPRO:

$$\bar{T}_w = T_s - qR_b^* - \frac{q}{2\pi k_s} g\left(\frac{t}{t_{sc}}, \frac{r_b}{H}\right) \tag{2}$$

where T_s (°C) is the undisturbed temperature of the subsurface, q (W m^{-1}) is the heat transfer rate per unit length of GHE, k_s is the subsurface thermal conductivity (W m^{-1} K^{-1}), t_{sc} (s; $t_{sc} = H^2/9\alpha_s$) is the time scale, and r_b (m) is the borehole radius. When normalizing the time with the time scale, α_s (m^2 s^{-1}) is the subsurface thermal diffusivity. The superposition principle takes into account temporal variations of the heat transfer rate calculated from monthly and peak building loads as well as the heat pump coefficient of performance affected by the water temperature leaving the GHEs. The long-term response of the GHE field is defined according to (Eskilson 1987) g (−) function, whose short-term response has been modified to consider the effect of the thermal mass of water contained in the system (Xu and Spitler 2006). When sizing and simulating GHEs, temperatures computed with GLHEPRO are affected by heat stored inside the water contained in the GHEs and all surface piping, an approach that is neglected with EED. Temperature changes due to short-term peak loads can be damped by the thermal mass of water, especially for coaxial GHEs that contain a greater amount of water than single and double U-pipe GHEs. An adequate approach for coaxial GHE design was to manually calculate the borehole thermal resistance affected by internal heat transfer with EED, and then specifying that value in GLHEPRO in an iterative manner for sizing and simulation, accounting for the thermal mass of water.

After sizing and simulating a ground-coupled heat pump system with various GHE configurations, the pumping power was calculated for U-pipe and coaxial GHEs and compared to one another. The Darcy-Weisbach equation was used to calculate head loss through piping under a worst-case scenario, i.e., when the viscosity of water is higher at lower temperatures. Those conditions are expected during each heating season when the GHEs are use to extract energy from the subsurface and the water temperature in the GHEs decreases. The pumping power was then calculated by multiplying the water flow rate and density, the gravitational acceleration and the head loss through all the GHEs.

3. RESULTS AND DISCUSSION

3.1 Sensitivity of design parameters to the borehole thermal resistance

The 3D borehole thermal resistance of coaxial GHEs with a borehole diameter of 152.4 and 203.2 mm (6 and 8 in.) was calculated with EED using average values of the parameters given in Table 1. GHE parameters were then varied one at a time to evaluate the impact of changing the water flow rate, the borehole length, the thermal conductivity of the backfilling, and the inner and outer pipe materials as well as the inner and outer pipe SDR (Figure 2). The minimum and maximum values given in Table 1were used to assess the sensitivity of each parameter. Increasing the flow rate reduced the borehole thermal resistance by up to 35%, with most variations when flow in the outer annulus changes from laminar to turbulent (Figure 2a). Increasing the borehole length from 50 to 250 m increased the borehole thermal resistance by about 2% (Figure 2b). This variation is due to the three-dimensional approach used with Equation 1 to evaluate internal heat transfer effects for the borehole thermal resistance and that, unlike methods limited by two dimensions, takes into account the borehole length at the numerator. Varying the thermal conductivity of the backfilling material changed the borehole thermal resistance by up to 41% (Figure 2c). Using a material having a thermal conductivity lower than that of HDPE for the inner pipe had a negligible impact of 0.6% on the borehole thermal resistance. However, increasing the outer pipe thermal conductivity reduced the borehole thermal resistance by up to 32% (Figures 2d, e). Similarly, changing the SDR of the inner pipe had little impact, while increasing the SDR of the outer pipe reduced the borehole thermal resistance by up to 32% (Figures 2f, g). In summary, calculations for coaxial GHEs indicate negligible advantages associated with insulating the inner pipe. Considerable advantages were obtained by increasing the water flow rate, the thermal conductivity of the outer pipe and the backfilling material as well as the outer pipe SDR. The main effect of increasing the outer pipe SDR is a reduce pipe thickness.

Borehole thermal resistance for single and double U-pipe GHEs was calculated with average parameters given in Table 1 for an SDR-11 pipe having a nominal diameter of 31.8 mm (1.25 in.). The spacing between the centers of the two sides of a U-pipe was 100 mm to be representative of a case with space clips that are commonly used. Both regular and thermally enhanced HDPE pipes of thermal conductivity equal to 0.4 and 0.7 $W\ m^{-1}\ K^{-1}$ were considered and used as reference cases. Obtained resistances were compared to those calculated for coaxial GHEs using average parameters (Table 1), except for the outer pipe thermal conductivity and SDRs that were increased. Coaxial GHEs had resistances up to 55% smaller than those of single U-pipe GHEs when increasing the outer pipe thermal conductivity or SDR (Figures 3a, b). Coaxial GHEs resistances were however higher than the double U-pipe GHEs considered in this study. The lowest resistance obtain with the coaxial configuration was about 22% higher than the resistance of the double U-pipe GHE made with thermally enhanced HDPE. In summary, it was necessary to improve the outer pipe material and dimensions of the coaxial GHE for its resistance to be smaller than the single U-pipe GHE, and it was not possible to have a resistance significantly smaller than the double U-pipe GHE.

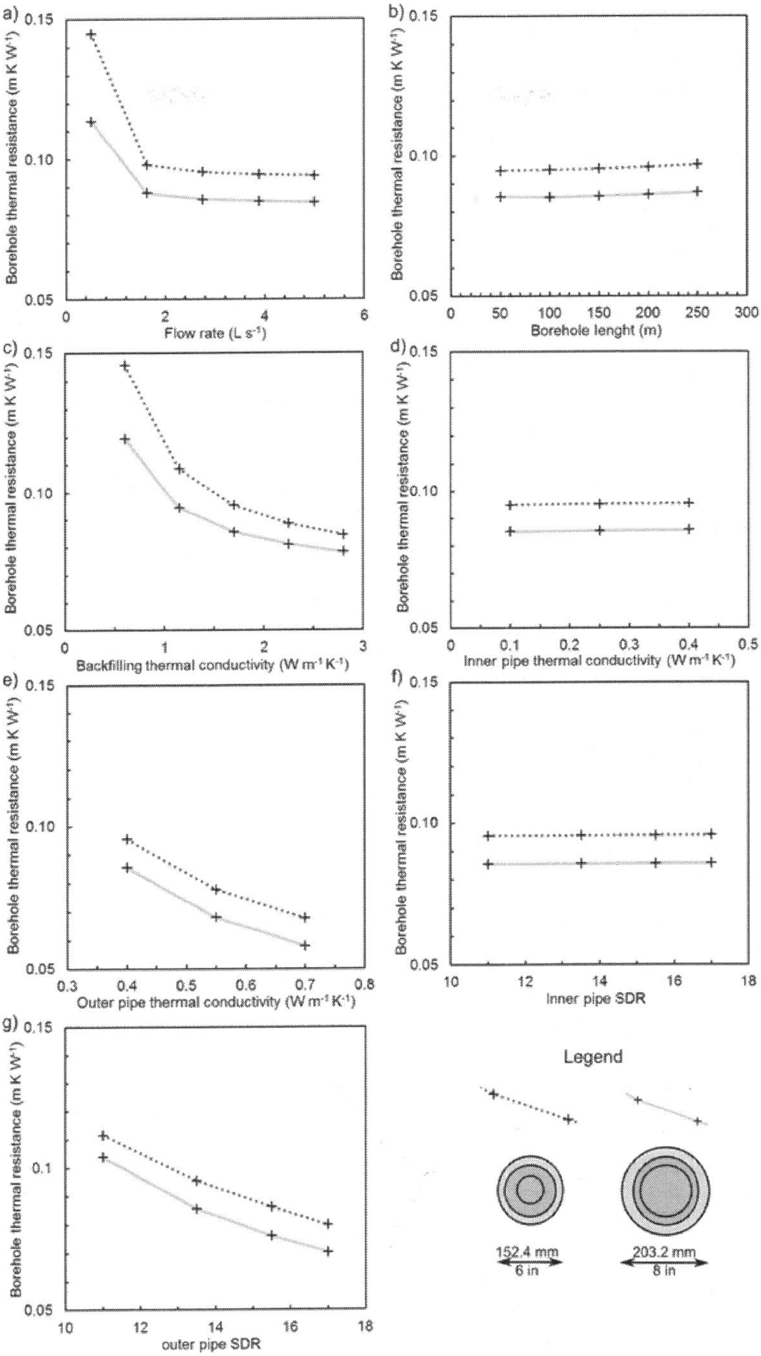

Figure 2. Borehole thermal resistance of coaxial GHEs. The GHE configuration belonging to each graphic line is shown in the legend with a cross section sketch of the pipes and borehole.

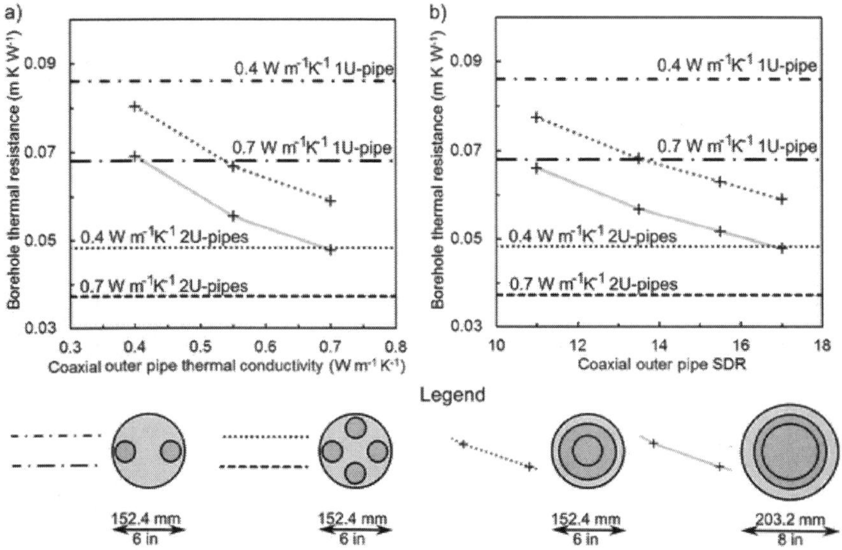

Figure 3. Comparison of borehole thermal resistance for U-pipe and coaxial GHEs. Properties of coaxial GHE only are varied on the *x* axis and are compared to U-pipe GHE resistances as reference cases. The GHE configuration belonging to each graphic line is shown in the legend with a cross section sketch of the pipes and borehole.

3.2 Thermal mass of water in the GHEs

The volume of water contained in a GHE that is 150 m long was calculated for four scenarios. The first two calculations were for a single and a double U-pipe GHE, whose pipe nominal diameter is 31.8 mm (1.25 in.). In the third and fourth cases, the volume of water was calculated for coaxial GHEs, where the outer and inner pipes have nominal diameters of 101.6 and 50.8 mm (4 and 2 in.) and 152.4 and 101.6 mm (6 and 4 in.), respectively. The water volume was then multiplied by the volumetric heat capacity of water (4.2 MJ m^{-3} K^{-1}) to determine the amount of energy needed to increase the temperature of water in the GHE by 1 K. When comparing the U-pipe versus the coaxial cases, the amount of energy was shown to be 6 to 28 times smaller (Figure 4). The increase of heat storage capacity associated with coaxial GHEs is expected to positively impact sizing calculations performed with GLHEPRO, where the fluid factor has been set to 1 for all cases presented below. The fluid factor in GLHEPRO is the ratio of the total fluid in the system to the fluid in the GHE. Since the amount of water in surface piping is expected to be similar, a value equal to 1 was used to compare each case, which neglected the water in the near surface piping connecting all of the GHEs together.

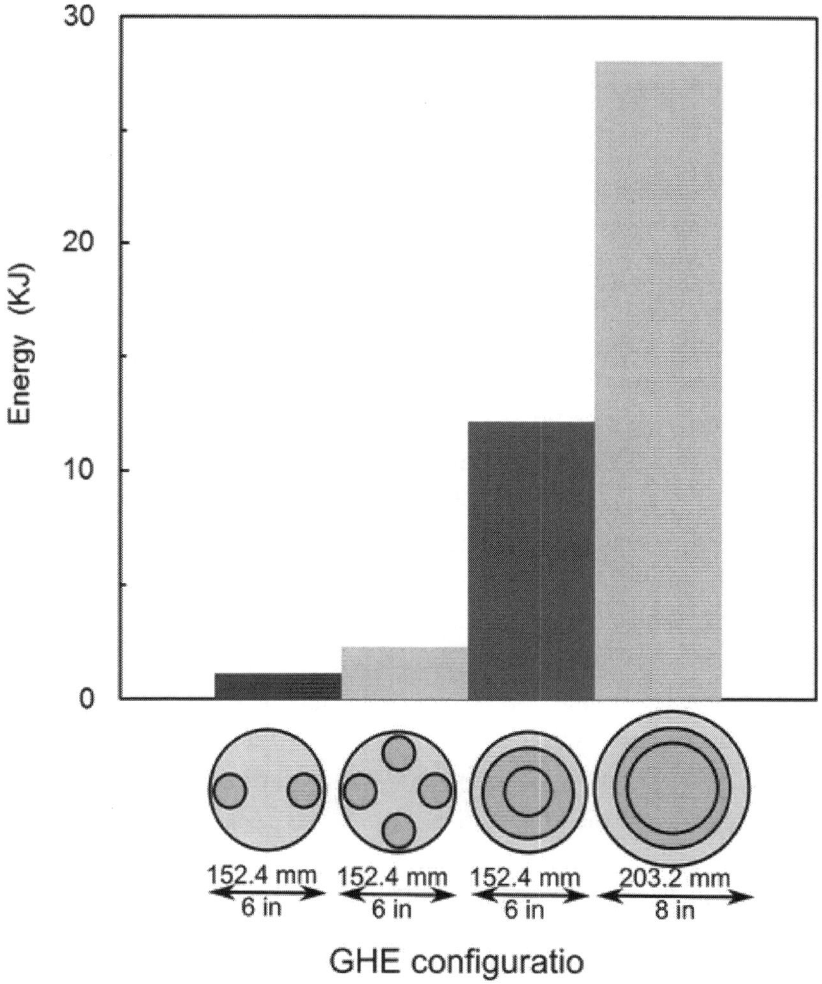

Figure 4. Energy required to increase the water temperature inside GHEs by 1 K. The GHE configuration belonging to each graphic line is shown in the legend with a cross section sketch of the pipes and borehole.

3.3 Sizing and simulation of a ground-coupled heat pump system

A synthetic building load profile was generated over a full year with a sinusoidal function for the sizing and simulation of a fictive ground-coupled heat pump system. The function was used to generate the fictitious hourly loads Q_{bl} (W; Bernier 2000):

$$Q_{bl} = A + B \sin\left(270 + \frac{h_y}{8,760} 360\right) \times \sin\left(\frac{h_d}{24} 180\right) \tag{3}$$

used as inputs for the sizing calculations. In Equation 3, A and B are constants that have been set to 50 and 100 kW, respectively. The arguments of the sin functions are in degrees, and h_y and h_d are the hours of the year (0 to 8,760) and day (0 to 24), respectively. The resulting load profile was cooling dominated, with a peak cooling load of −150 kW occurring in July and a peak heating load of 50 kW occurring in January (Figure 5). The choice of negative sign for cooling loads is to express heat rejected toward the subsurface. Daily peak loads were maintained for approximately 3 h. This choice of load profile is arbitrary and intended to require more cooling since cooling dominated buildings appear to be better suited for coaxial GHEs using pure water for the heat carrier fluid. The purpose was to investigate conditions where coaxial GHE could potentially be successful. Such scenario was also advantageous for double U-pipe GHEs that can be operated at a low flow rate when using pure water for the heat carrier fluid. Cooling-dominated profiles are additionally common to commercial buildings with internal gains found in Northeast America. The system was consequently sized according to the peak cooling loads and water temperature in GHEs remained above freezing. Antifreeze solutions such as propylene glycol used in GHEs operating below freezing conditions for heating dominated buildings increase the fluid viscosity and the flow rate needed for turbulence in coaxial GHEs. Designing a system with coaxial GHEs for a heating-dominated building is not impossible. For example, a calcium chloride solution of low viscosity with intermediate plate heat exchangers may be used but can be more complicated. This aspect is handled more easily with single U-pipe GHEs having a smaller pipe cross section and where fluid turbulence is reached at reasonable flow rates.

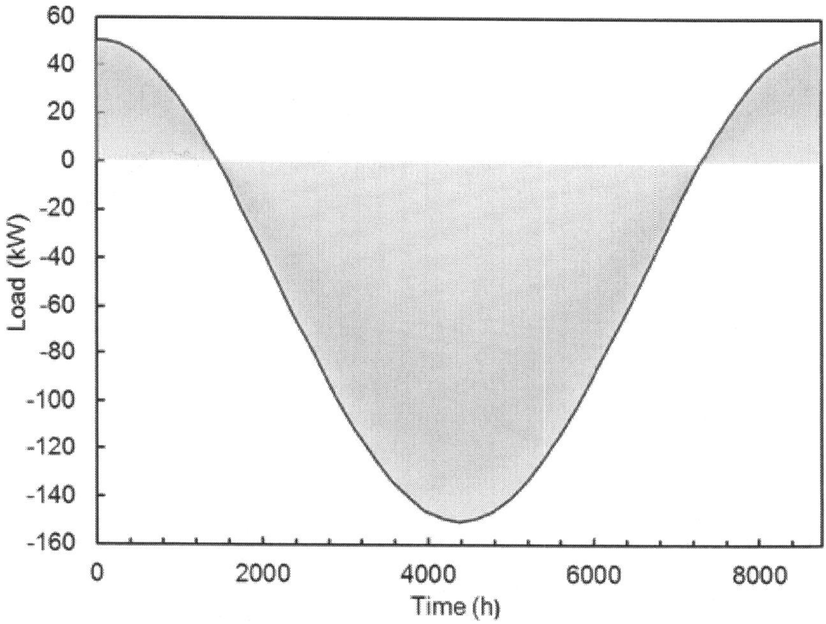

Figure 5. Synthetic load profile used for design calculations.

GHE fields for all sizing and simulation cases were rectangular. Boreholes were added when greater GHE length was needed, and the distance between each borehole was kept at 10 m to minimize thermal interaction between GHEs. The target depth of the boreholes for cases with single U-pipe GHEs was approximately 150 m, whereas those for cases with double U-pipe and coaxial GHEs were from 180 to 250 m. A greater depth is necessary with double U-pipe and coaxial GHEs for a system to have fewer boreholes such that the total system flow rate remains reasonably low. This ensures that the flow rate of individual GHEs is high and turbulent, even in the larger pipe of the coaxial GHEs. Balancing flow rates to minimize total system flow and maximize individual GHE flow is a key for designing coaxial GHEs that can be achieved by drilling to greater depths. Six cases with different U-pipe and coaxial GHEs were selected for the sizing calculations (Table 2).

The maximum water temperature chosen to enter the heat pump system during the peak cooling condition was 35°C. For the water temperature to fulfill this criterion during a 10-year period, sizing was initially performed when the full building loads were applied to the GHE field. A second round of sizing calculations was carried out using the same duration for a hybrid system that encloses a cooling tower with 55 kW of cooling capacity, approximately one third of the peak cooling load.

Table 2. GHE configuration and sizing calculation results for the synthetic load profile

GHE configuration						
Borehole diameter - mm (in.)	152.4 (6)	152.4 (6)	152.4 (6)	152.4 (6)	152.4 (6)	203.2 (8)
GHE configuration	1 U-pipe	1 U-pipe	2 U-pipe	2 U-pipe	Coaxial	Coaxial
Pipe nominal diameter - mm (in.)	31.8 (1.25)	31.8 (1.25)	31.8 (1.25)	31.8 (1.25)	101.6 (4) outer	152.4 (6) outer
					50.8 (2) inner	101.6 (4) inner
Pipe SDR	11	11	11	11	17 outer	17 outer
					11 inner	11 inner
Pipe thermal conductivity - W m^{-1} K^{-1}	0.4	0.7	0.4	0.7	0.7 outer	0.7 outer
					0.4 inner	0.4 inner
Grout thermal conductivity - W m^{-1} K^{-1}	1.7	1.7	1.7	1.7	1.7	1.7
Sizing calculation results when applying the full building loads						
Total flow rate - L s^{-1}	7	7	7	7	9.6	8
Borehole thermal resistance - m K W^{-1}	0.0955	0.0777	0.0563	0.0443	0.0734	0.0630
GHE grid	4 × 4	4 × 4	3 × 4	2 × 5	3 × 4	2 × 4
Individual GHE length - m	159	151	181	203	194	246
Total GHE length - m	2,544	2,416	2,172	2,030	2,328	1,968
Total water volume in GHE - m^3	4.62	4.39	7.89	7.37	16.43	27.48
Pumping power for peak conditions - W	246	234	138	213	130	31
Borehole length reduction - %	—	5	15	20	9	23
Sizing calculation results for a hybrid system with a 55-kW cooling tower						
Total flow rate - L s^{-1}	4.6	4.6	4.6	4.6	6	6
Borehole thermal resistance - m K W^{-1}	0.0948	0.0768	0.0547	0.0442	0.0701	0.0587
GHE grid	2 × 5	2 × 5	2 × 4	2 × 3	2 × 3	2 × 3
Individual GHE length - m	146	137	158	199	222	198
Total GHE length - m	1,460	1,370	1,264	1,194	1,332	1,188
Total water volume in GHE - m^3	2.65	2.49	4.59	4.37	9.40	16.59
Pumping power for peak conditions - W	160	152	77	160	137	19
Borehole length reduction - %	-	6	13	18	9	19

The sizing calculations indicated that reducing borehole thermal resistance and increasing water volume for the coaxial GHEs helped to decrease the total borehole length by 9% to 23% (Table 2). This compares to a 5% to 20% bore length reduction when using thermally enhanced HDPE with a single U-pipe, double U-pipe, or a combination of both. The bore length reduction for the single U-pipe GHEs with thermally enhanced HDPE can be more than 8% when the subsurface thermal conductivity is 4.5 W m^{-1} K^{-1} or higher (Raymond et al. 2011a). The subsurface thermal conductivity at the borehole interface, considered when using the multipole method (Claesson and Hellström 2011) to calculate the borehole thermal resistance of U-pipe GHEs, has a small effect on the resistance. This is due to the non-uniform temperature distribution at the borehole interface of U-pipe GHEs (Lamarche et al. 2010; Raymond et al. 2011b). A uniform temperature can be assumed for coaxial GHEs at the borehole interface in the case of homogenous subsurface because of the GHE axial symmetry. Calculation of the borehole thermal resistance for the coaxial GHE is therefore independent of the subsurface thermal conductivity. A constant value of 2.5 W m^{-1} K^{-1} was consequently used for the subsurface thermal conductivity in the current study. The coaxial GHEs with a borehole diameter of 203.2 mm (8 in.) were the most efficient configuration to reduce the total borehole length. The total system flow rate had to be increased for the coaxial GHEs to have a low borehole thermal resistance. Although higher flow rates were used, the pumping power associated with the coaxial GHEs remained lower because of the larger flow channels. The maximum temperature that entered the heat pump occurred at month 115 for the design cases using coaxial GHEs with a borehole diameter of 203.2 mm (8 in.; Figure 6). Other design cases showed a similar temperature evolution of the water in the GHEs and are not included on the graph to avoid repetition.

4. CONCLUSIONS

Design calculations performed for a fictive ground-coupled heat pump system using a synthetic load profile demonstrated that coaxial GHEs can reduce the total borehole length required for the system to fulfill a building's energy needs. Boreholes of coaxial GHEs studied in this manuscript had a diameter of 152.4 and 203.2 mm (6 and 8 in.) and have been compared with single and double U-pipe GHEs with a borehole diameter of 152.4 mm (6 in.). The parameters that needed to be improved for the coaxial GHEs to result in less total borehole length were the water flow rate as well as the outer pipe thermal conductivity and dimensions. It was helpful to select a thermally enhanced pipe made of HDPE with a thermal conductivity of 0.7 W m^{-1} K^{-1} and with a standard dimension ratio equal to 17 to form the outer flow channel of coaxial GHEs. Calculations suggested that at a high flow rate, the borehole thermal resistance of the proposed coaxial GHEs can be below 0.05 m K W^{-1}, which is significantly below single U-pipe GHEs but remains above double U-pipe GHEs. Another possible configuration to reduce the borehole thermal resistance is a coaxial GHE made with a flexible liner (Acuña and Palm 2013), which was shown to have a measured resistance that is 40% smaller than the lowest resistance of coaxial GHE considered in this study. Insulating the inner pipe with materials of thermal conductivity lower than that of HDPE had a negligible effect on the borehole thermal resistance reported in this study. Thermal short-circuiting between the inner and outer flow channel was not a major concern for the studied GHEs with a large diameter, which can be different for GHEs of small diameter (Zanchini et al. 2010). The volume of water in the proposed coaxial GHEs was up to 28 times greater than the volume for U-pipe GHEs, which is a second factor explaining the borehole length reduction. The large coaxial GHEs with a high heat storage capacity provided more bore length reduction than the double U-pipe GHEs even though the borehole thermal resistance was higher.

Sizing calculations considered the internal borehole thermal resistance (Hellström 1991) and the thermal mass of water (Xu and Spitler 2006) of the coaxial GHEs and enthusiastically revealed borehole length reductions of up to 23%. However, technical barriers still have to be addressed for those reductions to provide savings on installation costs. The coaxial GHEs with a borehole diameter of 203.2 mm (8 in.) showed most advantages associated with borehole thermal resistance and heat storage capacity. Commercial drilling capacity available in North America may need to accommodate larger boreholes for coaxial GHEs to be installed at competitive cost. Assembling HDPE pipes with a large diameter can complicate the installation process because large pipes are commonly shipped in sections to be joined with fusion tools in the field. The double U-pipe GHEs were shown to provide slightly smaller but similar advantages for the total borehole length reduction (up to 20%), especially when considering thermally enhanced pipe and deeper boreholes. While efficient technology to install coaxial GHEs may not be available at the moment, installation of double U-pipe GHE can be achieved with current tools and expertise. The drilling depth is an important factor for balancing flow rate of both double U-pipe and coaxial GHEs. Determining the depth at which this practice remains economical would require further research. In any cases, the thermally enhanced pipe was shown to be an asset for the alternative GHE configurations studied.

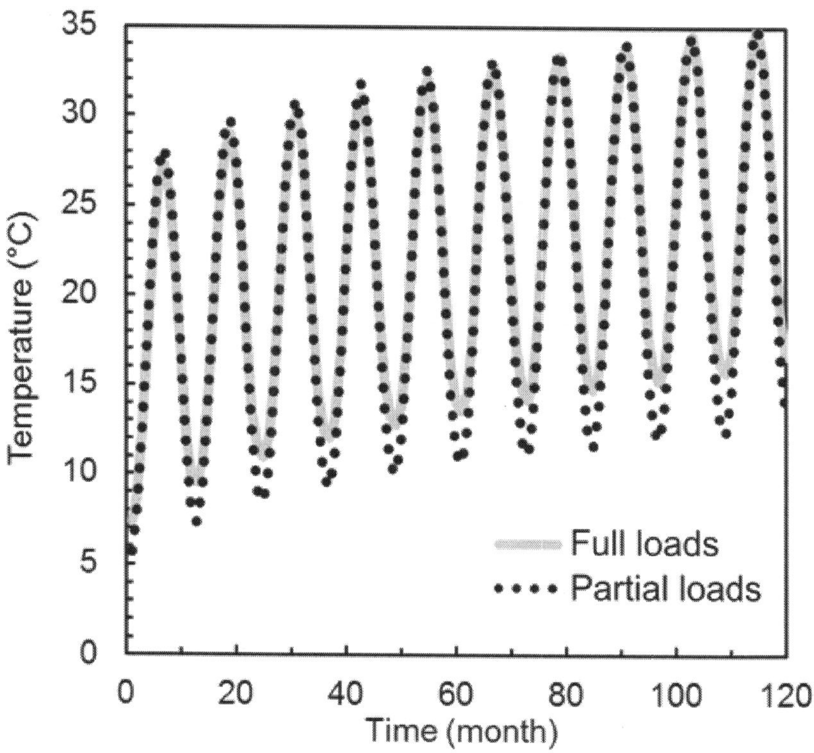

Figure 6. Maximum fluid temperature entering the heat pump for a system using coaxial GHEs. Borehole diameter is 203.2 mm (8 in.).

ACKNOWLEDGEMENTS

Professor Jeffrey Spitler at Oklahoma State University is kindly acknowledged for providing the program GLHEPRO to conduct the work presented in this manuscript. The comments of two anonymous reviewers were appreciated to improve this manuscript.

COMPETING INTERESTS

The first author of this manuscript has no competing interests. The second and third authors of this manuscript are working for Versaprofiles that produce thermally enhanced pipes for ground-coupled heat pump systems.

AUTHORS' CONTRIBUTIONS

The researcher JR performed the calculations and wrote the manuscript under the support and guidance of pipe industry specialists SM and LN to make results useful for practitioners. All authors read and approved the final manuscript.

REFERENCES

1. Acuña J, Mogensen P, Palm B (2011) Distributed thermal response tests on a multi-pipe coaxial borehole heat exchanger. HVAC & R Research 17:1012–1029

2. Acuña J, Palm B (2013) Distributed thermal response tests on pipe-in-pipe borehole heat exchangers. Appl Energy 109:312–320

3. Allan ML, Kavanaugh SP (1999) Thermal conductivity of cementitious grouts and impact on heat exchange length design for ground source heat pumps. HVAC & R Research 5:87–98

4. Beier RA, Acuña J, Mogensen P, Palm B (2013) Borehole resistance and vertical temperature profiles in coaxial borehole heat exchangers. Appl Energy 102:665–675

5. Bernier M (2000) A review of the cylindrical heat source method for the design and analysis of vertical ground-coupled heat pump systems. In: Proceedings of the fourth International conference on heat pumps in cold climates, Caneta Research Inc., Aylmer, Canada.

6. Borinaga-Treviño R, Pascual-Muñoz P, Castro-Fresno D, Blanco-Fernandez E (2013a) Borehole thermal response and thermal resistance of four different grouting materials measured with a TRT. Appl Therm Eng 53:13–20

7. Borinaga-Treviño R, Pascual-Muñoz P, Castro-Fresno D, Del Coz-Díaz JJ (2013b) Study of different grouting materials used in vertical geothermal closed-loop heat exchangers. Appl Therm Eng 50:159–167

8. Canadian GeoExchange Coalition (2010) The state of the Canadian geothermal heat pump industry 2010 - industry survey and market analysis. Public Report, Canadian GeoExchange Coalition, Montreal, Canada, http://www.geo-exchange.ca/en/UserAttachments/article64_Industry%20Survey%202010_FINAL_E.pdf. Accessed November 3rd, 2014

9. Carlson SW (2000) GSHP bore field performance comparisons of standard and thermally enhanced grout. ASHRAE Trans 106:442–446

10. Claesson J, Hellström G (2011) Multipole method to calculate borehole thermal resistances in a borehole heat exchanger. HVAC & R Research 17:895–911

11. Eskilson P (1987) Thermal analysis of heat extraction boreholes. Ph. D Thesis, Lund Institute of Technology, Department of mathematical physics, Lund, Sweden

12. Gustafsson A-M, Westerlund L (2011) Heat extraction thermal response test in a groundwater-filled borehole heat exchanger - Investigation of the borehole thermal resistance. Renew Energy 36:2388–2394

13. Hellström G (1991) Ground heat storage; thermal analysis of duct storage systems. Ph. D. thesis, University of Lund, Department of mathematical physics, Lund, Sweden

14. Kavanaugh SP, Allan ML (1999) Testing of thermally enhanced cement ground heat exchangers grout. ASHRAE Trans 105:446–450

15. Lamarche L, Kajl S, Beauchamp B (2010) A review of methods to evaluate borehole thermal resistances in geothermal heat-pump systems. Geothermics 39:187–200

16. Raymond J, Frenette M, Léger A, Magni E, Therrien R (2011a) Numerical modeling of thermally enhanced pipe performances in vertical ground heat exchangers. ASHRAE Trans 117:899–907

17. Raymond J, Therrien R, Gosselin L (2011b) Borehole temperature evolution during thermal response tests. Geothermics 40:69–78

18. Raymond J, Therrien R, Gosselin L, Lefebvre R (2011c) A review of thermal response test analysis using pumping test concepts. Ground Water 49:932–945

19. Xu X, Spitler J D (2006) Modeling of vertical ground loop heat exchangers with variable convective resistance and thermal mass of the fluid. In: Proceedings of the 10th international conference on thermal energy storage, Ecostock, Ponoma NJ: The Richard Stockton College of New Jersey

20. Zanchini E, Lazzari S, Priarone A (2010) Improving the thermal performance of coaxial borehole heat exchangers. Energy 35:657–666

CHAPTER 5

Fouling and Fouling Mitigation on Heat Exchanger Surfaces

S. N. Kazi

[1] Department of Mechanical and Materials Engineering, Faculty of Engineering, University of Malaya, Kuala Lumpur, Malaysia

1. INTRODUCTION

Heating or cooling of one medium by another medium is performed in a heat exchanger along with heat dissipation from surfaces of the equipment. In course of time during operation, the equipment receives deposition (Fouling) which retards heat exchanging capability of the equipment along with enhanced pressure loss and extended pumping power. Thus accumulation of undesired substances on a surface is defined as fouling. Occurrence of fouling is observed in natural as well as synthetic systems. In the present context undesired deposits on the heat exchanger surfaces are referred to fouling. With the development of fouling the heat exchanger may deteriorate to the extent that it must be withdrawn from service for cleaning or replacement.

The overall design of heat exchanger may significantly be influenced by fouling, use of material, process parameters, and continuous service in the system or process stream are all deliberately influenced by fouling phenomena. Preventive measures of fouling are highly encouraged as it keeps the service of heat exchanger for a longer time. However many mitigation techniques of fouling are harsh to the environment. A technique involving chemicals and means benign to the environment is the most desired approach and it could elongate the cleaning interval. On the other hand unique and effective arrangements may be required to facilitate satisfactory performances between cleaning schedules. As a result fouling causes huge economic loss due to its impact on initial cost on heat exchanging operation, operating cost, mitigation measures and performance. The present study focused on fouling phenomena, fouling models, environment of fouling, consideration of heat exchanger fouling in design and mitigation of fouling.

2. FOULING

Fouling is the resultant effect of deposition and removal of deposits on a heat exchanger surface. The process of fouling could be represented by the equation (1).

$$\frac{dm_f}{dt} = \dot{m}_d - \dot{m}_r \tag{1}$$

Where dmf, m•d and m·r are net deposition rates, deposition and removal rates respectively.

Various deposition and removal processes for a typical system could be predicted as shown in Figure 1. The processes occur simultaneously and depend on the operating conditions. Usually removal rates increase with increasing amounts of deposit whereas deposition rates are independent of the amount of deposit but do depend on the changes caused by deposits such as increase in flow velocity and surface roughness. In the application of constant wall temperature or constant heat transfer coefficient boundary conditions, the interface temperature decreases as deposits build up which reduces the deposition rate.

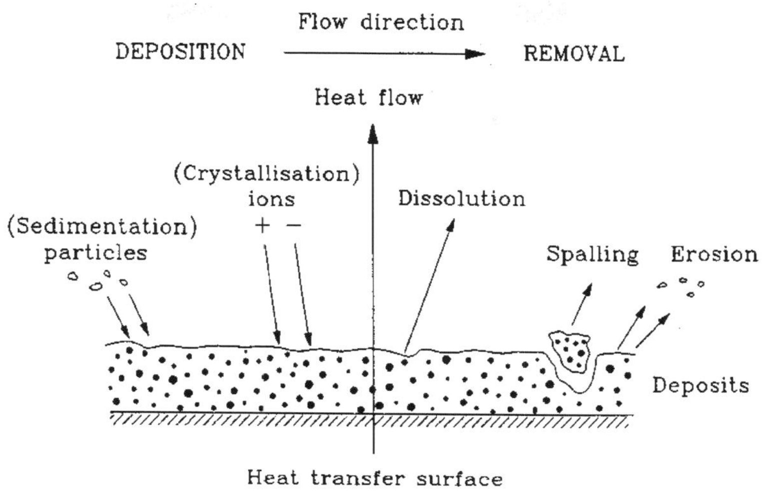

Figure 1. Various deposition and removal processes during fouling.

Initiation period or time delay in heat exchanger fouling is considered the time when there is no deposition for some time after a clean heat exchanger has been brought into operation. Figure 2illustrates this in detail. The initial growth of deposit can cause the heat transfer coefficient to increase rather than decrease resulting in a fouling resistance due to changing flow characteristics near the wall. At the initial stage the deposit penetrates the viscous sub-layer, the resulting turbulence increases the film heat transfer coefficient at the solid/liquid interface by changing flow characteristics near the wall. This increase in heat transfer coefficient may overcome the thermal resistance offered by the deposits and the net heat transfer coefficient may increase.

Several authors have reported negative fouling resistances [1, 2]. This process continues until the additional heat transfer resistance overcomes the advantage of increased turbulence. The time period from the beginning of the fouling process until the fouling resistance again becomes zero is called roughness delay time [3]. The time period from the beginning, when the formation of stable crystalline nuclei and their concretion to a compact fouling layer takes place is also called as induction period, which is in fact the roughness delay time and it ends up with the increase of fouling resistance above zero level.

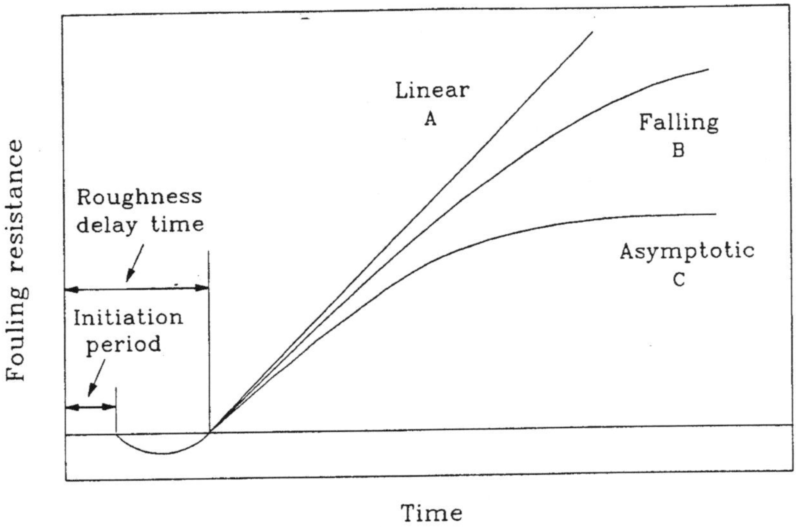

Figure 2. Typical fouling curves.

The initiation period and the roughness delay time for particulate fouling are very small [4] in comparison to the fairly long delay time for crystallization fouling [5]. After the roughness delay time, the fouling curve can be classified into three categories, (a) Linear, (b) Falling, and asymptotic, as illustrated in Figure 2.

The linear fouling curve is obtained for very strong deposits where removal is negligible or in case where the removal rate is constant (and deposition is faster than removal). The falling rate curve is obtained from decrease in deposition and deposits with lower mechanical strength. The combined effect with time causes the net deposition or fouling rate to fall. Asymptotic fouling curve has been most commonly reported for different types of fouling. The removal rate increases with time for weak deposits and can eventually become equal to the deposition rate. The net rate is then zero as depicted in Figure 2.

Linear fouling curves have been presented by many authors for crystallization fouling [6-8]. However, there is some doubt as to whether the fouling rate may remain linear for a long time. For the constant heat flux situation, the net driving force may decrease with fouling. The increase in flow velocity due to the reduced cross-sectional area with deposit formation can increase the removal rate and the linear rate may change to a falling rate or even level-off completely [9]. Asymptotic behavior for crystallization fouling has reported by various authors [10, 5, 11-12]. Cooper et al. [13] found asymptotic behavior for calcium phosphate fouling (with some particulate fouling from suspended solids). For particulate fouling, asymptotic behavior is attained because particles do not adhere strongly to the wall and can be removed easily [4, 14].

A fouling process that follows a linear rate for constant heat flux can have falling or even asymptotic behaviour for constant temperature difference. The interface temperature decreases with deposit formation because of the extra resistance offered by deposit layer and enhanced flow velocities as flow passages are partially blocked by deposits. Thus the thermal boundary conditions can result in different fouling curves which may give wrong impressions about the actual fouling mechanism.

2.1. Categories of Fouling

Fouling can be categorised a number of different ways. These are (1) heat transfer service, (2) type of service fluid and (3) application. Most fouling situations are virtually unique. Fouling [15] can be classified into the following categories: (i) particulate, (ii) Precipitation, (iii) corrosion, (iv) biofouling and (v) chemical reaction.

2.2. Particulate Fouling

Particulate fouling is evolved by the accumulation of solid particles suspended in the process stream onto the heat transfer surface. Heavy particles settle on a horizontal surface due to gravity and fine particles settle onto heat transfer surfaces at different inclinations due to suction force or other mechanisms. Unburned fuels or ashes deposition on boiler tubes, dust deposition on air cooled condensers etc. are examples of particulate fouling.

2.3. Precipitation Fouling (Sedimentation Fouling)

This kind of fouling is also called crystallization fouling. Dissolved inorganic salts are normally present in fluid used in heat exchanger. There is a maximum amount of the salt (saturated) which can be dissolved in this fluid. During heating or cooling supersaturation occurs in the dissolved inorganic salts. The inverse solubility salts such as calcium and magnesium sulphate, carbonate, silicate, etc. have less solubility in warm water up to a certain temperature than in cold water. This may occur when the process condition inside the heat exchanger is different from condition at the entrance. A stream on a wall at a temperature above that of corresponding saturation temperature for the dissolved salts allows crystal formation on the surface. Normally crystallization starts at especially active points – nucleation sites – such as scratches and pits and often after induction period spread to cover the entire surface. This type of fouling is strong and adherent and requires vigorous mechanical or chemical treatment to be removed [16]. Fouling rate increases with the increase of salt concentration or surface temperature. These are often found in heat exchangers of process industries, boilers, evaporators etc.

2.4. Chemical Reaction Fouling

This type of fouling occurs when the depositions are formed as a result of chemical reaction resulting to produce a solid phase at or near the surface. In the present case carbonaceous material deposits due to thermal gradation of the components of a process stream on hot heat transfer surface. This type of fouling is often extremely tenacious and need special measure to clean off the deposits from heat exchanger surfaces to provide them satisfactory operation life [16].

2.5. Corrosion Fouling

This type of fouling is also caused by some chemical reaction but it is different from chemical reaction fouling. This fouling is a reactant and it is consumed. In this case, the surface reacts with the fluid and become corroded [15]. The corrosion products can foul the surface provided it is not dissolved in the solution after formation. pH value of the solution is one of the controlling parameter. Such as, presence of sulfur in fuel can cause corrosion in gas and oil fired boilers. Corrosion is often more prone in the liquid side of the heat exchanger. In some cases the product of corrosion may be swept away to downstream of a process loop and cause deposition on surfaces there.

2.6. Accumulation of Biological Fouling

On a heat transfer surface the growth of biological materials results in biofouling. In this case biological micro and macro organisms are stick to the heat transfer surface. When microorganisms (e.g., algae, bacteria, molds etc.) and their products grow they form microbial fouling. Seaweeds, waterweeds, barnacles develop microbial fouling. These fouling may occur simultaneously. The growth of attached organisms is one of the common problems [15] in heat exchanger operation. Food processing industries, power plant condensers using seawater, etc. are experiencing biofouling.

2.7. Fouling Process

Fouling is a complex phenomenon due to involvement of a large number of variables. From a fundamental point of view the fouling mechanism follows certain stages in developing on a surface [17]. These are: Initiation, transport, attachment, removal and aging.

2.8. Initiation

Surface is conditioned in the initiation period. The initial delay induction period is influenced by the materials surface temperature, material, surface finish, roughness and surface coating. With the increase of degree of supersaturation with respect to the heat transfer surface temperature or increase of surface temperature the induction period decreases. During the induction period, nuclei for crystallization of deposit are also formed for biological growth. This period can take a long time, may be several weeks or a few minutes or even seconds.

The delay period decreases with increasing temperature in chemical reaction fouling due to the acceleration of induction reactions. If the initial period decreases with increasing surface temperature, crystallization fouling would be changed [18]. With the increase of surface roughness the delay period tends to decrease [19]. Additional sites are developed by the roughness projections, which promotes crystallization while grooves provide regions for particulate deposition.

2.9. Transport

In this part, fouling substances from the bulk fluid are transported to the heat transfer surface across the boundary layer. This is dependent on the physical properties of the system and concentration difference between the bulk and the surface fluid interface. Transport is accomplished by a number of phenomena including diffusion, sedimentation and thermophoresis [20, 21]. The local deposition flux md on a surface can be expressed by equation (2).

$$\dot{m}_d = h_D\left(C_b - C_s\right) \qquad (2)$$

Where, Cb and Cs are reactant concentration in the bulk fluid and that in the fluid adjacent to the heat transfer surface where as hD is the convective mass transfer coefficient. From Sherwood number (Sh=hDd/D), hD could be evaluated. Sherwood number is dependent on the flow and the geometric parameters.

The phenomenon of transportation of a particulate matter in a fluid due to gravity on a horizontal or inclined surface is known as sedimentation. This is playing a vital roll where particles are heavy and fluid velocities are low.

2.10. Attachment

At this stage, the deposits are adhered to the surface and among itself. Salt ions approaching to the surface are attracted to it due to electro-magnetic forces and adhere to the surface to form nucleation and gradually it grows with time to form a fouling layer. Thus forces acting on the particles as they approach the surface are impotent in determining attachment. Properties of the materials, such as size, density and surface conditions are dominating the attachment phenomenon.

2.11. Removal

There is competition between removal and deposition of the foulants, up to the steady growth of the deposition on the surface. Shear forces at the interface between the fluid and deposited fouling are responsible for removal. The velocity gradients at the surface, the viscosity of the fluid and surface roughness are guiding the shear forces. Removal from the surface is performed through the mechanism of dissolution, erosion and spalling.

2.12. Aging

With the commencement of deposition ageing starts. During the ageing, there may be transformation of crystal to improve or decrease the deposition strength with time. During aging the mechanical properties of the deposit can change due to changes in the crystal or chemical structure. Alteration of the chemical composition of the deposit by a chemical reaction may change its mechanical strength. On the other hand the biofouling layer may become weak due to corrosion at the surface by slow poisoning of microorganisms.

2.13. Change in Deposition Thickness With Time

Figure 3 is showing the growth rate of deposit on the surface [15]. Region A: fouling is initiated in the induction period. Region B: a steady deposit growth on the surface. The rate of removal of deposit was increased when the rates of deposition gradually retards. Region C: in this region the rate of removal and deposition seems equal and the thickness of deposition remains constant.

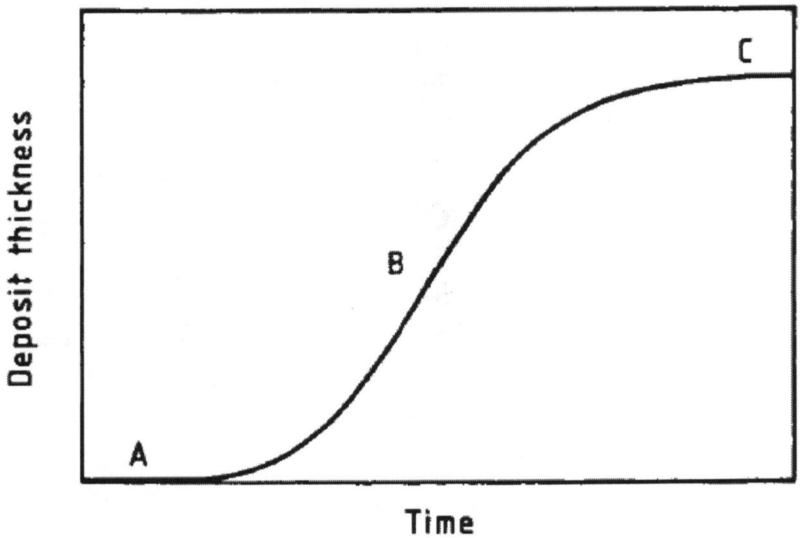

Figure 3. Change in deposition thickness with time.

2.14. Composite Fouling

Some of the common salts causes fouling are $CaSO_4$, $CaCO_3$ and $Mg(OH)_2$, and SiO_2. Solubility, crystal structure and strength have impact on composite scale formation in fouling. Therefore, composite fouling needs more attention and further research [17].

3. EFFECTS OF FOULING

Fouling phenomena imposes retardation on heat transfer and augmentation of frictional pressure drop which degrades the effectiveness of a heat exchanger. Some basic design aspects of heat exchangers along with mitigation of fouling are discussed in the present chapter.

3.1. Effect Of Fouling on Heat Exchanger Design

A fixed value of fouling resistance could be assigned during the design stage although fouling is time dependent phenomenon. The cleaning schedule and operating parameters of the heat exchanger is dependent on the design fouling factor. Depending on application some heat exchangers require frequent cleaning whereas some need rear cleaning. Fouling rate is a dominating factor in designing a particular heat exchanger.

Fouling allowance: Provisions are during the design stage once fouling is anticipated. Different approaches are used to provide an allowance for fouling resistance. They all result into an excess heat transfer surface area. Updated methods include, specifying the fouling resistances, the cleanliness factor, or the percentage over surface.

A fouling resistance is prescribed on each side of the surface where fouling is anticipated. A lower overall heat transfer coefficient is resulted. To achieve the specified heat transfer, excess surface area is provided. Until the specified value of the fouling resistance is reached, the performance of the heat exchanger will be satisfactory. Depending on this fact, maintenance schedule could be planned to avoid unprecedented shut down for cleaning. Tubular Exchanger Manufacturers Association (TEMA) [22] is referenced source of fouling factors used in the design of heat exchangers. Plant data, proprietary research data, personal and company experience etc. are other sources of fouling resistance data could be used in design.

Minimize Fouling by considering Design Features: Extent of fouling could be minimized by good design practice. Direct contact heat exchangers are considered where excess fouling is desired. In general a fouling prone fluid stream should be placed on the tube side as cleaning is easier. Generally higher fluid velocity and lower tube wall temperature retard fouling accumulation. Velocity of 1.8 m/s is a widely accepted figure for tube side flow of a heat exchanger. Heat Exchangers, operating over dew point for acid vapor and above freezing for fluids containing waxes prevent corrosion and freezing fouling. Fouling deposits are always found heavy in the region of low velocity at the vicinity of baffles in the shell side of the shell and tube heat exchangers.

Design features to facilitate fouling control: Full elimination of fouling may not be possible by good design practice alone. So, heat exchangers require cleaning at certain intervals. On-line cleaning can be employed to control fouling by extending cleaning cycle. Continuous fouling can ensure minimized fouling allowance. At construction and installation phase of a plant on-line cleaning system could be installed at ease. A heat exchanger with removable head and straight tube would be easy to clean and maintain. Space and provision for removal and cleaning of tube bundles are required to be available. On site cleaning facilities are to be provided with options of keeping isolation valves and connection provisions for cleaning hoses which could lead to chemical cleaning.

Fouling and operation of heat exchangers: Provision of excess surface area in heat exchangers for curbing fouling may lead to operation problem and fouling build. Generally high heat transfer area enhances total heat transfer which raises the out let temperature. By changing process parameters such as flow, surface temperature leads to higher fouling.

Fouling control strategies: A number of strategies are applied for fouling control. In operating condition additives are added. On-line or off-line surface cleaning techniques are other options. To control fouling under different consequences are consolidated by some researchers as stated in Table 1[23].

Table 1. Various techniques adapted to control fouling.

On-line techniques	Off-line techniques
Use and control of appropriate additives: Inhibitors, Antiscalants, Dispursants, Acids, Air jet	Disassembly and manual cleaning: Lances: Liquid jet, Steam, Air jet. Mechanical Clearing: Drills, Scrapers
On-line cleaning: Sponge balls, Brushes, Sonic horns, Soot blowers, Chains and scrappers, Thermal shock, Air bumping	Chemical cleaning

Heat Exchanger with green additives: Many additives were developed for retardation of fouling but many of them found carcinogenic in nature. Now researchers are heading towards green additives. Chemistry and analysis are underway. Lab analysis and performances will be subsequently achieved. In near future users are looking for a breakthrough in this field.

3.2. Fouling Effect on Heat Transport

Mineral scales deposited on heat exchanger surfaces are a persistent and an expensive problem in process industries, cooling water systems, steam generation units, desalination by evaporation etc. and also house hold equipment. Precipitation of mineral salts as a scale on the surface of the conduit and cause obstruction of fluid flow, impedance of heat transfer, wear of metal parts, localized corrosion attack and unscheduled equipment shutdown.

The deposit layer provides an additional resistance to heat transfer. Generally, the thermal conductivity of the deposit layer is very low compared with that of the material of the heat exchanger which may result in a much higher thermal resistance than the wall or film resistances. The deposit layer also reduces the flow area, which increases the pressure drop. This problem is quite severe and is further enhanced by the rough surface of the deposit. Both effects reduce the heat exchanger performance significantly. Additional energy requirements in terms of more heating or pumping power can hamper the economics of the process.

In a circular tube, fouling builds on the inside or outside of the tube depending on the flowing fluid. Fouling adds an insulating cover to the heat transfer surface. The overall heat transfer coefficient for a smooth tubular heat exchanger under deposited conditions, U_f can be obtained by adding the inside and outside thermal resistances:

$$U_f = \frac{1}{A_o / A_i h_i + A_o R_{fi} / A_i + A_o \ln(d_o / d_i) / 2\pi k L + R_{fo} + 1/h_o} \quad (3)$$

where R_{fi} and R_{fo} represent resistances for the outer and inner surfaces of the tubes.

The thermal resistance due to fouling is evaluated generally based on experiments as difference in the overall specific resistances of the fouled and clean wall:

$$R_f = \left(\frac{1}{U_f} - \frac{1}{U_{cl}}\right) \quad (4)$$

Where, the overall heat transfer coefficient U_f can also be evaluated by using the rate equation:

$$U_f = \frac{\dot{Q}}{(A \times \Delta T_f)} \quad (5)$$

The heat flow rate ·Qf•and temperature difference ΔTf (the temperature difference between heated surface and the bulk liquid) are experimentally obtained. A is the exposed area of the heat exchanging surface to the liquid. The net rate of deposition of $CaSO_4.2H_2O$ on metal surface is estimated as $\frac{m}{t}$ where m is the total mass accumulation on a unit area and t refers to the amount of time the surface was exposed to the solution of the foulant.

Using the definition of heat transfer coefficient and fouling resistance, the equation (6) can be derived for constant heat duty.

$$\frac{A_{fouled}}{A_{clean}} = 1 + U_{clean} R_f \qquad (6)$$

The required excess heat transfer area usually becomes excessive due to the higher clean heat transfer coefficients. It is often recommended that the additional surface should not exceed 25 percent of the heat transfer surface requirement for clean operation.

3.3. Effect of Fouling on Pressure Drop

In heat exchangers pressure loss is considered more critical than loss in heat transfer due to fouling. Fouling results in a finite layer. Flow field, pressure drop are affected by the change in geometry of the flow passage. Thus in a tubular heat exchanger, the deposited layer roughens the surface, diminishes the inner and raises the outer dimension of the tubes. The inside diameter of the tube decreases and roughness of the tube increases due to fouling which, causes an increase in pressure drop. Pressure drop inside a tube of a heat exchanger under fouled and clean state can be correlated as follows:

$$\frac{\Delta P_f}{\Delta P_c} = \frac{f_f}{f_c}\left(\frac{d_c}{d_f}\right)\left(\frac{u_{mf}}{u_{mc}}\right)^2 \qquad (7)$$

Considering that the mass flow rates under clean and fouled conditions are the same, the mass flow rate can be represented as:

$$\dot{m} = \rho u_m A_{cr} \qquad (8)$$

Equation (7) thus becomes:

$$\frac{\Delta P_f}{\Delta P_c} = \frac{f_f}{f_c}\left(\frac{d_c}{d_f}\right)^5 \qquad (9)$$

The magnitude of df of scaled tube can be obtained from equation (10).

$$d_f = d_c \exp\left(-\frac{2k_c R_f}{d_c}\right) \tag{10}$$

The thickness tf of deposit layer can be obtained from:

$$t_f = 0.5d_c\left[1 - \exp\left(-\frac{2k_f R_f}{d_c}\right)\right] \tag{11}$$

For a known total fouling resistance, the tube diameter under fouled conditions can be evaluated on knowing the thermal conductivity of the deposits. Non-uniform thermal conductivity may result from the multi layers of fouling deposits. Approximate thermal conductivities of pure materials constituting fouling deposits are often used for estimation of thermal conductivity of the total deposits. Depending on situations the fouling layer is considered composed solely of one material. In some occasions to ease calculations ff is considered equal to fc.

4. CONDITIONS INFLUENCING FOULING

The conditions influencing fouling can be classified as: (A) operating parameters, (B) heat exchanger parameters, and (C) fluid properties. Among the operating parameters the important events which influencing fouling at a significant level are: (1) velocity, (2) surface temperature, and (3) bulk temperature.

Velocity influences fouling at a significant level. In diffusion controlled processes, increasing the fluid velocity causes more fouling [24]. In most cases, fouling decreases at higher fluid velocities [4, 13,25]. Increasing flow velocity increases the fluid shear stress which causes more removal. This results in lower fouling rates which resulting to lower fouling resistance. For weak deposits (particulate fouling), increasing the flow velocity may completely eliminate fouling. For stronger deposits, increasing the flow velocity beyond a particular point may not decrease fouling significantly [25]. For very strong deposits, increasing the flow velocity may not have any effect at all [6].

Surface temperature may increase, decrease or have no effect on fouling [26]. The rates of chemical reaction and inverse solubility crystallization increase with an increase in temperature. For inverse solubility salts, higher surface temperature increases fouling due to higher concentration gradients and higher reaction rate constants. In case of normal solubility salts cooling results in more fouling.

The bulk temperature also effects on increase of fouling rate. In inverse crystallisation, when precipitation happens in the fluid bulk, increasing the temperature increases the rate of crystal formation and hence deposition. Thus bulk temperature has effects on chemical reaction rate and polymerisation rate.

The important heat exchanger parameters are classified as: surface material, surface structure (roughness), heat exchanger type and geometry [27]. Surface material is considered seriously for corrosion fouling because of the potential to react and form corrosion products. Different materials have different catalytic action and may promote or reduce fouling for different processes. The initial fouling rate and scale formation depends significantly on the surface roughness. Junghahn [28] proved theoretically that the free energy change associated with crystal nuclei formation was much less on a rough surface than on a smooth surface. Rough surfaces result in higher deposition due to protected zones in the cavities or pits where flow velocities are very low.

According to Rankin and Adamson [29], it is not the rate of nucleation but the nuclei attachment which is strongly dependent on the surface roughness. Chandler [30] also observed similar results. In general the rough surface causes more fouling which reduces the delay time for all types of fouling. Surface roughness increases turbulence near the surface, which in turn increases the removal rate of fouling on the surface. Better performance occurred due to the increase in surface roughness with deposit formation and has been reported by some authors [1, 2]. Marriott [31] reiterated that mirror finished surfaces in heat exchangers are used to reduce fouling in practice.

5. HEAT EXCHANGER TYPE, GEOMETRY AND PROCESS FLUID INFLUENCING FOULING

Shell and tube heat exchangers are used most commonly but they are not particularly suitable for fouling conditions. Fouling can be reduced with special baffle and tube design. Several studies [32-35] have shown that finned tubes foul less than plain tubes. Non-uniform thermal expansion leads to lower deposit strength and hence less deposition. Freeman et al. [36] found that tubes with longitudinal grooves on the outside had less particulate fouling (by alumina particles) than the plain tubes.

Fluidised-bed heat exchangers are used in several applications to reduce or even eliminate fouling completely. Fluidised particles remove deposits from the heat transfer surface. They also enhance the heat transfer efficiency as they interrupt the viscous sub-layer. These heat exchangers have been used successfully to reduce fouling by hard, adhering silica deposits [37]. Graphite heat exchangers are also reported to have less fouling. Direct contact heat transfer may be another alternative to reduce fouling [38]. Properties of the process fluid such as the nature and concentration of the dissolved constituents or suspended particles, presence of any living organisms, solution pH etc. affect fouling significantly.

Excessively high over-concentration of solids in the evaporating liquid may lead to carry-over in the steam and cause fouling in process heat transfer equipment. Corrosion is very important on the steam side of process equipment. Water pH, over-concentration of treatment chemicals in evaporating liquids and dissolved gases (mainly oxygen and carbon dioxide) are very important contributors to corrosion fouling [39]. The presence of living organisms causes biological fouling and makes biofilms. This can sometimes enhance other fouling mechanisms too, as microbial deposits may trap suspended particles. They may also change the chemistry of water and can cause scaling or corrosion [39].

6. FOULING MODELS

A number of models have been proposed for different types of fouling. Analysis and model improvement is still progressing as there are difficulties due to the complex nature of deposit formation and lack of reproducible measurement of fouling resistance. Most of the models have been simplified with many assumptions [40] as stated below:

- Surface roughness is neglected.
- Change in surface roughness with deposit formation is also neglected.
- Only one type of fouling is usually considered.
- Changes in physical properties of the fluids are neglected in most of the cases.
- The fouling layer is assumed to be homogeneous.
- Changes in flow velocity with changing cross-sectional area due to fouling are usually neglected.
- The shape of deposits, e. g. crystals or particles is ignored.

It is also observed that few attempts have been made to model the initiation or roughness delay period. Almost all the models predict fouling (scaling) after the delay period. Some other notable parameters are neglected in modelling such as: (a) effect of simultaneous action of different fouling mechanisms, (b) equipment design, (c) surface parameters e.g. surface material and surface roughness, (d) increase in surface area with deposition, (e) properties of foulant stream, (f) nature of process, and the (g) fluctuations in operation.

Modelling is usually done taking into consideration only (a) flow velocity, (b) concentration, (c) wall and bulk temperature, and (d) time.

Watkinson and Martinez [11] developed a model, based on the fundamental material balance equation (1). For the deposition rate the following expression is adopted:

The deposition rate is expressed as shown in equation (12).

$$\frac{dx_f}{dt} = \frac{K_R}{\rho_f}(c_F - c_{Sa})^n \qquad (12)$$

For sparingly soluble salts with inverse solubility (e.g. $CaCO_3$), the deposition rate is controlled by the slow reaction rate and the constant of reaction rate KR that obeys the Arrhenius equation:

$$K_R = A_0 e^{\left(\frac{-E}{R_g T_f}\right)} \qquad (13)$$

with T_f as the interface temperature.

Kern and Seaton [43] recommend for the removal rate the equation:

$$\dot{m}_r = a_8 \tau_f x_f \qquad (14)$$

Where τ_f is the shear stress exerted by the liquid flow on the fouling film. Even though $CaCO_3$deposits are much stronger than the particulate deposits considered by Kern and Seaton [43] the removal rate was assumed to be directly proportional to deposit thickness, which may not be correct for all the cases.

Kern and Seaton [41] proposed a model for particulate fouling which takes into account removal or re-entrainment of deposits. The mathematical model is based on a general material balance equation (1). Deposition and removal rates act separately and combine into a net deposition rate. The rate of deposition is expressed as:

$$\dot{m}_d = a_9 c' w \qquad (15)$$

Where, 'c' is dirt concentration and w is constant weight flow of fluid. The removal rate is roughly proportional to the total depth of dirt deposited on the heat transfer surface as stated below.

$$\dot{m}_r = a_{10}\tau_f x_f \tag{16}$$

Combining the equations for deposition and removal rates (15) and (16) with the material balance equation (1), the fouling resistance expression is obtained:

$$R_f = \overset{*}{R}_f(1 - e^{-\theta t}) \tag{17}$$

where θ is a time constant and R∗f is the asymptotic value of the fouling resistance. For these also the following equations are obtained.

$$\overset{*}{R}_f = \frac{a_9 c' w}{a_{10}\lambda_f \tau_f} \tag{18}$$

$$\theta = a_{10}\tau_f \tag{19}$$

Here, λ_f is the thermal conductivity of the deposits, a9 and a10 are proportionality constants. This model predicts asymptotic fouling behaviour with R∗f being the fouling resistance after an infinite time of operation. According to this model, no matter what the conditions, i.e. type of fluid, heat exchanger surface, temperature driving force, an asymptotic fouling value will be obtained sooner or later with removal rates becoming equal to deposition rates.

7. COST IMPOSED DUE TO FOULING

An additional cost is imposed by fouling of heat transfer equipment in industries. Few studies have been undertaken to determine the fouling related costs in industry. Fouling costs can generally be divided into four major categories, such as (1) increased capital expenditure, (2) energy costs, (3) maintenance costs, (4) cost of production loss and (v) extra environmental management cost.

Table 2. Estimated fouling costs incurred in some countries.

Country	Fouling costs US $ million	GNP (1984) US $ million	Fouling costs % of GNP
USA (1982)	3860-7000	3634000	0.12-0.22
	8000-10000		0.28-0.35
Japan	3062	1225000	0.25
West Germany	1533	613000	0.25
UK (1978)	700-930	285000	0.20-0.33
Australia	260	173000	0.15
New Zealand	35	23000	0.15
Total Industrial World	26850	13429000	0.20

The heat transfer area of a heat exchanger is kept exaggerated to compensate retardation imposed by fouling. Oversized pumps and fans are selected to compensate design over-surfacing the enhanced pressure loss from reduction in the flow area.

In some occasions standby heat exchangers are kept in process design in order to ensure uninterrupted operation while a fouled heat exchanger is taken under cleaning maintenance. In-situ cleaning in some cases are recommended while chemical cleaning is preferred for others. All together, cost of cleaning, cleaning equipment, chemicals all are imposing extra to the capital cost of the plant.

Muller-Steinhagen [37] reported that total annual costs for highly industrialised countries such as the United States and the United Kingdom are about 0.25 percent of the countries gross national product (GNP). Even for a relatively less industrialised country like New Zealand, the total fouling costs are around 0.15 percent of its GNP. Muller-Steinhagen [37] has summarised the total fouling costs for various countries based on 1984 in Table 2.

8. FOULING MITIGATION

Gilmour [42] reported that the degradation of heat transfer performance due to fouling in shell and tube heat exchangers occurs mainly due to poor shell-side design. In recent years numerous methods have been developed to control fouling. These methods can be classified as: (1)

chemical methods, (2) mechanical methods and (3) changing the phase of the solution. By adding foreign chemicals in a solution, reduction of fouling is achieved by chemical methods of fouling mitigation. Chemical additives developed by many companies have been extensively used to mitigate fouling in the industrial sector. Various additives can be used to prevent scaling [43-44]. Bott [45] specified that the additives used act in different ways, such as (a) sequestering agents, (b) threshold agents, (c) crystal modifiers and (d) dispersants. Some of the common water additives are EDTA (sequestering agent), polyphosphates and polyphosphonates (threshold agents) and polycarboxylic acid and its derivatives (sequestering and threshold treatment). Sequestering agents such as EDTA complex strongly with the scaling cations such as Ca^{++}, Mg^{++}, and Cu^{++} in exchange with Na^{+}, thus preventing scaling as well as removing any scale formed previously. They are used effectively as antiscalants in boiler feed water treatment. Troup and Richardson [46] claimed that their use is uneconomical when hardness levels are high.

Polyphosphates and polyphosphonates as threshold agents are also used to reduce scaling in boilers and cooling water systems. Bott [45] said that they prevent the formation of nuclei thus preventing the crystallisation and mitigate fouling. Very small quantities of these agents are effective in reducing scaling from supersaturated salt solutions.

Crystal modifying agents (e.g. Polycarboxylic acid) distort the crystal habit and inhibit the formation of large crystals. The distorted crystals do not settle on the heat transfer surface, they remain suspended in the bulk solution. If their concentration increases beyond a certain limit, particulate fouling may take place. This is prevented either by using techniques to minimise particulate fouling or using dispersing agents along with crystal modifying agents.

Though crystallisation fouling may not be prevented completely using additives, the resulting crystalline deposits are different from those formed in the absence of any additives. The layer looses its strength and can be removed easily. By controlling pH, crystallisation fouling can furthermore be minimised. The solubility of deposit forming components usually increases with decreasing pH. In many water treatment plants, sulphuric acid is added to maintain a pH between 6.5 and 7.5 [47]. In this case, addition of corrosion inhibitors may also be required which may enhance fouling again.

Seeding is used commercially to reduce crystallisation fouling. This method involves addition of seeds to the scaling fluid. Crystallisation takes place preferentially on these seeds rather than on the heat transfer surface. Calcium sulphate seeds are generally used to avoid calcium sulphate scaling [48-49]. These seeds need not be of the crystallising material, but they should have similar crystallographic properties, i. e. atomic agreement and lattice spacing [50].

To mitigate particulate fouling by chemical means, dispersants are used to reduce the surface tension of deposits. It helps in disintegrating the suspended particles into smaller fragments that do not settle so readily.

Addition of certain chemicals can slow down or terminate chemical reactions. Dispersants are very helpful in keeping the foulants away from the surface. Some particles such as corrosion products may act as catalysts. Chemical reaction fouling could be suppressed by reducing the number of these particles. Corrosion inhibitors (chromates and polyphosphates) can be used to reduce corrosion fouling [47]. Usually a passivating oxide layer is desired to prevent corrosion of the surface. Corrosion fouling may promote other fouling mechanisms e. g. higher roughness of the corroded surface may enhance crystallisation fouling. The corrosion products may act as catalysts and promote chemical reaction fouling and also augments particulate fouling by depositing on the heat transfer surface.

Mitigation of fouling by chemical methods has several drawbacks. Fouling and corrosion inhibitors usually contain considerable amount of chlorine, bromine, chromium, zinc etc. Therefore, their concentration has to be monitored carefully. Treatment of fluid released from the plant to natural waterways is necessary to prevent harmful effects. Higher concentrations can be used in closed systems but overdosing may have negative effects and some components may precipitate. Using different additives at the same time may result in dangerous chemical reactions. Some additives have limited life and some degrade with time and loose activity.

Pritchard [51] has broadly classified mechanical methods into two categories according to their ways of action. (1) Brute force methods such as high-pressure jets, lances, drills etc. (2) Mild methods such as brushes and sponge balls. Muller-Steinhagen [37] has reported that several mechanical methods have been developed in recent years. The following mechanisms predict the modern methods:

- Breakage of deposits during brief overheating due to differential thermal expansions of heat transfer surface and deposits,
- Mechanical vibration of the heat transfer surfaces,
- Acoustical vibration of the surface,
- Increased shear stress at the fluid/deposit interface, and
- Reduced stickiness of the heat transfer surface.

Most liquid-side fouling mitigation techniques have been developed for the tube-side of shell and tube heat exchangers. The relevant techniques include:

- increase in flow velocity,
- reversal of flow direction,
- heat transfer surface such as, surface roughness and surface materials,
- fluidised bed heat exchangers,
- pulsating flow,
- turbulence promoters, and
- transport of cleaning devices through tubes.

The deposits which are not strongly adhere to the surface can be removed by increasing the flow velocity. Muller-Steinhagen and Midis [4] reported that alumina deposits were removed completely when the flow velocity was increased for a short period of time after a fouling run. At higher flow velocity, the wall shear stress increases and causes more removal of deposits from the surface.

At a regular interval of time, the reversal of flow direction on the heat transfer surface could be another effective method of reducing fouling. This technique needs several modifications in the existing set-up. Muller-Steinhagen [37] stated that mitigation of fouling by increasing the flow velocity could be more effective than reversal of flow direction.

Surface material and surface roughness play an important role on fouling mitigation. Thus lowering the surface roughness retards the adhesion of deposits and the number of nuclei growth sites. Lower deposition rate also experienced with lowering surface energy of the material of heat exchanger. Using inert particles is an effective way of reducing or even eliminating fouling completely as practiced in fluidized bed heat exchangers. Pulsating flow in heat exchangers is a strategy to increase the level of turbulence [52-58]. Where, as a matter of fact heat transfer coefficient increases with the enhancement of deposit removal. Higher heat transfer reduces fouling by reducing the interface temperature which is beneficial for certain fouling mechanism such as crystallization fouling of inverse solubility salts. The higher level of turbulence augments the deposit removal rate.

Fracture of deposits by fatigue is enhanced by higher turbulence due to pulsation resulting to increase of removal rate. Generally the deposition rate of fouling phenomena [3] depends on the thickness of viscous and thermal sub-layers. Muller-Steinhagen [37] reported that by inserting turbulence promoters inside tubes or by using tube corrugations, the heat transfer coefficient can be increased by a factor of 2 to 15 by reducing the thickness of average thermal boundary layer. Turbulence promoters may reduce both the crystallisation and reaction fouling. Muller-Steinhagen [37] informed that particulate fouling will be enhanced if particulate or fibrous material already exists in the solution.

Middis [10] also reported fouling mitigation by adding natural fibre in the supersaturated solutions of concentration 3.6 g/L $CaSO_4$. He observed that the rate of $CaSO_4$ fouling on heated metal tube surface decreases with the increase of fibre concentration in the fouling solution. Kazi [59] also got similar results by adding different types and concentrations of natural fibre in supersaturated solutions of $CaSO_4$.

Some novel methods which do not fall under well reported categories, such as magnetic or electric treatment are also available in the market to reduce fouling. Usually magnetic treatment is carried out by inserting permanent magnets in a pipe before the heat exchanger.

Parkinson and Price [60] have reported significant reduction in fouling by the magnetic treatment as it helps in precipitating the salts. These salts stay suspended in the bulk liquid and are removed later. On the other hand Hasson and Bramson [61] informed that there is no effect of magnetic treatment at all on fouling. They observed that magnetic treatment neither decreased nor increased the rate of scaling. The nature of the deposits also remained unchanged. Bernadin and Chan [62] have also reported no influence of magnetic treatment on silica fouling. Muller-Steinhagen [37] has stated that magnetic mitigation devices in some cases actually increased fouling. Thus from the available information no conclusion can be made about the influence of the magnetic field on the scaling process.

9. CLEANING OF HEAT EXCHANGERS

A decrease in the performance of a heat exchanger beyond acceptable level requires cleaning. In some applications, the cleaning can be done on line to maintain acceptable performance without interruption of operation. At other times, off-line cleaning must be used.

Garrett-Price et al. [27] presented some cleaning approaches for fouled heat exchangers. They specified on-line cleaning generally utilises a mechanical method designed for only tube side and requires no disassembly. In some applications flow reversal is required. Chemical feed can also be used as an on-line cleaning technique but may upset the rest of the liquid service loop.

On-line mechanical cleaning techniques are also in practice. On line tube side cleaning techniques are the sponge-ball and brush systems. The advantage of on-line cleaning is the continuity of service of the exchanger and the hope that no cleaning-mandated downtime will occur. The principal disadvantage is the added cost of a new heat exchanger installation or the large cost of retrofits. Furthermore there is no assurance that all tubes are being cleaned sufficiently.

Off-line chemical cleaning is a technique that is used very frequently to clean exchangers. Some refineries and chemical plants have their own cleaning facilities for dipping bundles or re-circulating cleaning solutions. In general, this type of cleaning is designed to dissolve the deposit by means of a chemical reaction with the cleaning fluid. The advantages of chemical cleaning approach include the cleaning of difficult-to-reach areas. Often in mechanical cleaning, there is incomplete cleaning due to regions that are difficult to reach with the cleaning tools. There is no mechanical damage to the bundle from chemical cleaning, although there is a possibility of corrosion damage due to a reaction of the tube material with the cleaning fluid. This potential problem may be overcome through proper flushing of the unit. Disadvantages of off-line chemical cleaning include corrosion damage potential, handling of hazardous chemicals, use of a complex procedure.

Off-line mechanical cleaning is a frequently used procedure. The approach is to abrade or scrap away the deposit by some mechanical means. The method includes high-pressure water, steam, lances and water guns. In off-line mechanical cleaning there are some advantages such as excellent cleaning of each tube is possible, good removal potential of very tenacious deposits. Disadvantages include the inability to clean U-tube bundles successfully, usual disassembly problem and the great labour needed.

Frenier and Steven [63] describe general methods for cleaning heat exchanger equipment, including both mechanical and chemical procedures. They have given guidelines for selecting between chemical and mechanical cleaning, and among the various types of chemical cleaning processes. They stated that water-based fluids can transport and deposit a wide variety of minerals, and corrosion products form due to the reaction of the aqueous fluids with the metals of construction. Hydrocarbon and petrochemical fluids transport and deposit a variety of organic scales. Common inorganic scale forming compound includes various iron oxides, hardness deposits (carbonates and silicates). They stated that the entire cleaning situation must be considered when choosing between mechanical and chemical cleaning, as well as the specific technique within the general category. The general categories of mechanical cleaning are abrasive, abrasive hydraulic, hydraulic and thermal [64].

Frenier and Barber [63] stated that, for chemical cleaning of the heat exchanger tubes, it is very beneficial to obtain a sample of the deposit so that its composition can be determined. Based on the chemical analysis of the deposit, an optimal treatment plan can be developed and the best solvent selected. They have classified the deposits generically, as organic (process-side) or inorganic (water-side).

They stated that the process side deposits may range from light hydrocarbon to polymers and generally they are similar to the fluids from which they originate. They mentioned that the general categories of solvents for process side scales include aqueous detergent solutions, true organic solvents and emulsions. Aqueous detergent formulations always contain a surfactant-type component. In addition they can contain alkaline agents, such as sodium hydroxide, sodium silicate, or sodium phosphate. Builder molecules such as ethylenediaminetetraacetic acid (EDTA) suppress the effects of hard water, and coupling agents such as glycol ethers, improve the dissolution of some organic deposits.

Detergent formulations are effective only for removing the lighter deposits. Refinery fluids, aeromatics and terpenes are used to dissolve the organic deposits. N-methyl-2-pyrrolidinone also is a very effective polar solvent with low toxicity characteristics. They reiterated that the effectiveness of the application depends greatly on proper application conditions, such as flow rate and temperature. Combination of surfactants, organic solvents and water emulsions are good cleaning agents. Emulsions with an organic outer phase are particularly useful for cleaning large vessels. Oily rust deposits having both organic and inorganic compositions can be removed by acidic emulsions combining an acid and an organic solvent.

Water-side deposits usually contain minerals, such as iron oxides (corrosion products), hardness (Ca and Mg carbonates) and silica, in individual cases other minerals can also be found. The solvents for removing inorganic deposits usually contain mineral acids, organic acids or chelating agents.

Mineral acids used in chemical cleaning include hydrochloric acid (HCl), hydrofluoric acid (HF), sulphuric acid (H_2SO_4), phosphoric acid (H_3PO_4), nitric acid (HNO_3) and sulfamic acid (H_2NSO_3H). Hydrochloric acid is the most common and most versatile mineral acid. It is used on virtually all types of industrial process equipment at strengths from 5 percent to 28 percent (5-10 percent is the most usual range). It can be inhibited at temperatures up to about 180 F. HCl will dissolve carbonates, phosphates, most sulphates, ferrous sulphide, iron oxides and copper oxides. By using with appropriate additives, fluoride deposits, copper and silica can also be removed from surfaces with inhibited HCl. HCl is corrosive, so it has restricted use. HCl is not used to clean series 300 SS, free-machining alloys, magnesium, zinc, aluminium, cadmium, or galvanised steel because of the potential for generalised or localised attack. It is not desirable to contact the fouled metal with a strong mineral acid, because of the danger of damage to the equipment during or after cleaning. An alternative solvent family consists of aqueous solutions of chelating agents and organic acids with pH values of about 2 to 12.

Citric acid was one of the first organic acids used in industrial cleaning operations [65]. For removing iron oxide from steel surfaces, citric acid and a mixture of formic and citric acid could be used [66]. The mixture could hold more iron in solution than either of the acids alone could do. Ammonium citrate and sodium citrate solvents are currently used to clean a wide variety of heat transfer equipment, including boilers and various types of service water systems. The advantage of citric acid formulation is their low toxicity and ready biodegradability. EDTA is a versatile chemical that forms metal ion complexes with higher equilibrium constants than citric acid. As a result chemical cleaning solvents with pH values from 4.5 to about 9.2 have been formulated that can remove Fe and Cu, as well as Ca, Ni and Cr. The major advantage of the EDTA solvents is that they are much more aggressive than citric salts for removing very heavy iron oxide deposits especially if they contain copper. The disadvantage includes high cost per pound of metal removed and low biodegradability.

All of the chelating agents are also organic acids. Eberhard and Rosene [67] taught the use of solvents consisting of formic acid or citric acid for cleaning nondrainable tubes in super heaters. Reich [66] used a mixture of formic acid and citric acid to a proportion of 3:1, to remove iron oxide deposits. The advantage of these mixtures is that they avoid the precipitation of solids that formed in pure formic or citric acid solutions. Formulations of formic acid with hydroxyacetic acid and citric acid with hydroxyacetic acid can be used as a cleaning agent. Bipan [3] used acetic acid of concentration 3 percent to remove $CaSO_4.2H_2O$ deposit on plate type SS heat exchangers. He said that with the increase in acid solution temperature the removal efficiency increases. Similar results were obtained by Kazi [68]. It reveals that a complete and systematic study of fouling on different metal surfaces and their mitigation by additives have been required to be done along with study of introducing a benign to environment technique for chemical cleaning of fouling deposits.

REFERENCES

1. T. R. Bott, J. S. Gudmundsson, 1978 "Rippled Silica deposits in Heat Exchanger Tubes". 6th International Heat Transfer Conference.

2. B. D. Crittenden, E. M. H. Khater, 1987 "Fouling From Vaporising Kerosine". Journal of Heat Transfer, 109: 583589 .

3. B. Bansal, 1994 "Crystallisation Fouling in Plate Heat Exchangers", PhD thesis, Department of Chemical and Materials Engineering, The University of Auckland: Auckland, New Zealand.

4. H. M. Muller-Steinhagen, J. Middis, 1989 "Particulate Fouling in Plate Heat Exchangers". Heat Transfer Engineering, 10(4): 3036 .

5. M. Bohnet, 1987 "Fouling of Heat Transfer Surfaces". Chemical Engineering Technology, 10: 113125 .

6. R. B. Ritter, 1983 "Crystallisation Fouling Studies". Journal of Heat Transfer, 105: 374378 .

7. B. J. Reitzer, 1964 "Rate of Scale Formation in Tubular Heat Exchangers". I & EC Process Design and Development, 3(4): 345348 .

8. D. Hasson, 1962 “Rate of Decrease of Heat Transfer Due to Scale Deposition”. DECHEMA Monograph, 47: 233282 .

9. T. R. Bott, R. Walker, 1971 1971 “Fouling in Heat Transfer Equipment”. The Chemical Engineer, : 391395 .

10. J. Middis, 1994 “Heat Transfer and Pressure Drop For Flowing Wood Pulp Fibre Suspensions”, PhD thesis, Chemical and Materials Engineering. The University of Auckland: Auckland, New Zealand.

11. A. P. Watkinson, O. Martinez, 1975 „Scaling of Heat Exchanger Tubes by Calcium Carbonate”. Journal of Heat Transfer, 504508 .

12. W. Augustin, 1992 „Verkrustung (Fouling) Von Warmeubertragungsflachen“, in Institut fur Verfahrens- und Kerntechnik. Technische Universitat Braunschweig: Germany.

13. A. Cooper, J. W. Suitor, J. D. Usher, 1980 “Cooling Water Fouling in Plate Heat Exchangers”. Heat Transfer Engineering, 1(3): 5055 .

14. H. M. Muller-Steinhagen, F. Reif, N. Epstein, A. P. Watkinson, 1988 “Influence of Operating Conditions on Particulate Fouling”. The Canadian Journal of Chemical Engineering, 66: 4250 .

15. T. R. Bott, 1995 „Fouling of Heat Exchangers“. Elsevier Science & Technology Books. 529.

16. K. J. Bell, A. C. Mueller, 2001 „Wolverine Heat Transfer Data book II.“, Wolverine Tube, Inc.

17. N. Epstein, 1983 “Heat Exchanger Theory and Practice”, in: J. Taborek, G. Hewitt (eds.) heat exchangers in Heat Exchanger Theory and Practice, McGraw-Hill,

18. N. Epstein, 1983 „Thinking about Heat transfer fouling: a 5 x 5 matrix“. heat Transfer Engineering, 4: 4346 .

19. N. Epstein, 1981 “Fouling in heat exchangers”. In Low Reynolds Number Flow Heat Exchangers, S. Kakac, R. K. Shah, and A. E. Bergles (eds.). Hemisphere, New York.

20. E. F. C. Somerscales, J. G. Knudsen, (eds, 1981 “Fouling of Heat Transfer Equipment. Hemisphere”, New York.

21. L. F. Melo, T. R. Bott, C. A. Bernardo, (eds, 1988 “Fouling Science and Technology”. Kluwer, Dordrecht.

22. “Tubular Exchanger Manufacturers” 1988 Association Standards of the Tubular Exchanger Manufacturers Association 7th ed., New York,

23. J. M. Chenoweth, 1988 “General design of heat exchangers for fouling conditions”. In Fouling Science and Technology, L. F. Melo, T. R. Bott, and C. A. Bernardo (eds.), 477494 . Kluwer, Dordrecht.

24. M. Brusilovsky, J. Borden, D. Hasson, 1992 “Flux Decline due to Gypsum Precipitation on RO Membranes”. Desalination, 86: 187222 .

25. G. Walker, 1982 “Degradation of Performance”, Industrial Heat Exchangers- A Basic Guide. Hemisphere Publishing Corporation. 213272 .

26. J. S. Gudmundson, 1981 “Particulate Fouling”, in Fouling of Heat Transfer Equipment, E.F.C. Somerscales and J.G. Knudsen, Editors. Hemisphere Publishing Corporation. 357387 .

27. B. A. Garrett-Price, S. A. Smith, R. L. Watts, J. G. Knudsen, W. J. Marner, J. W. Suitor, 1985 „Overview of Fouling, Fouling of Heat Exchangers- Characteristics, Costs, Prevention, Control, and Removal”. Noyes Publications: New Jersey. 920 .

28. L. Junghahn, 1964 „Methoden Zum Herabsetzen oder Verhindern der Krustenbildung“. Chemie Ingenieur Technik, 36: 6067 .

29. B. H. Rankin, W. L. Adamson, 1973 “Scale Formation as Related to Evaporator Surface Conditions”. Desalination, 13: 6387 .

30. J. L. Chandler, 1964 “The Effect of Supersaturation and Flow Conditions on the Initiation of Scale Formation”. Transactions of Institution of Chemical Engineers, 42: T24T34 .

31. J. Marriott, 1971 "Where and How to Use Plate Heat Exchangers". Chemical Engineering, 78(8): 127134 .

32. J. G. Knudsen, H. K. Mc Cluer, 1959 "Hard Water Scaling of Finned Tubes at Moderate Temperatures". Chem. Eng. Progress Symp. series, 55(9): 14 .

33. D. L. Katz, J. G. Knudsen, G. Balekjian, S. S. Grover, 1954 "Fouling of Heat Exchangers". Petroleum Refiner, 33(4): 121125 .

34. W. O. Webber, 1960 "Under Fouling Conditions Finned Tubes can Save Money". Chemical Engineering, 676: 149152

35. R. Sheikholeslami, A. P. Watkinson, 1986 "Scaling of Plain and Externally Finned Tubes". Journal of Heat Transfer, 108: 147152 .

36. W. B. Freeman, J. Middis, H. M. Muller-Steinhagen, 1990 "Influence of Augmented Surfaces and of Surface Finish on Particulate Fouling in Double Pipe Heat exchangers". Chemical Engineering Processing, 27: 111 .

37. H. M. Muller-Steinhagen, 1993 "Fouling: The Ultimate Challenge for Heat Exchanger Design". The sixth International Symposium on Transport Phenomena in Thermal Engineering. Seoul, Korea.

38. H. M. Muller-Steinhagen, 1988 "Introduction to Heat Exchanger Fouling". Proceedings of Fouling in Heat Exchangers. The University of Auckland, Centre for Continuing Education, Auckland, New Zealand.

39. B. A. Garrett-Price, S. A. Smith, R. L. Watts, J. G. Knudsen, W. J. Marner, J. W. Suitor, 1985 "Generic Industrial Fouling", in Fouling of Heat Exchangers- Characteristics, Costs, Prevention, Control, and Removal. Noyes Publications. 2137 .

40. J. D. D. R. S. Pinheiro, 1986 "Fouling of Heat Transfer surfaces", Heat Exchanger Source Book, J.W. Palen, Editor. Hemisphere Publishing Corporation. 721744 .

41. D. Q. Kern, R. E. Seaton, 1959 "A Theoretical Analysis of Thermal surface Fouling". Brit. Chem. Eng., 4(5): 258262 .

42. C. H. Gilmour, 1965 “No Fooling- No Fouling”. Chem. Eng. Progr., 61(7): 4954 .

43. A. Harris, A. Marshall, 1981 “The Evaluation of Scale Control Additives”, Conference on Progress in the Prevention of Fouling in Industrial Plant. University of Nottingham.

44. A. S. Krisher, 1978 “Raw Water Treatment in the CPI”. Chemical Engineering, 7998 .

45. T. R. Bott, 1981 “The Fouling of Heat Exchangers”. DSIR, Wellington, New Zealand.,

46. D. H. Troup, J. A. Richardson, 1978 “Scale Nucleation on a Heat Transfer Surface and its prevention”. Chemical Engineering Communications, 2: 167180 .

47. H. M. Muller-Steinhagen, 1993 “Fouling of Heat Transfer Surfaces”. VDI Heat Atlas, English Edition, VDI- Verlag GmbH, OC1OC22 .

48. R. J. Gainey, C. A. Thorp, Cadwallader, 1963 “Calcium Sulphate Seeding Prevents Calcium sulphate Scaling”. Industrial and Engineering Chemistry, 55(3): 3943 .

49. R. Rautenbach, R. Habbe, 1991 “Seeding Technique for Zero-Discharge Processes”, Adaption to Electrodialysis. Desalination, 84: 153161 .

50. M. Telkes, 1952 “Nucleation of Supersaturated Inorganic Salt Solutions”. Industrial and Engineering Chemistry, 44(6): 13081310 .

51. A. M. Pritchard, 1988 “Cleaning of Fouled Surfaces: A Discussion, in Fouling Science and Technology”, NATO ASI Series, Series E: Applied Science, Melo, L. F., Bott, T. R. and Bernardo, C. A., Editors. 721726 .

52. R. H. Keil, 1971 “Enhancement of Heat Transfer by Flow Pulsation”. Industrial Engineering Chemistry: Process Design and Development, 10(4): 473478 .

53. J. C. Ludlow, D. J. Kirwan, J. L. Gainer, 1980 “Heat Transfer with Pulsating Flow”. Chemical Engineering Communications, 7: 211218.

54. O. E. Karamercan, J. L. Gainer, Effect. The, Pulsations. of, Heat. on, Transfer, Industrial Engineering Chemistry Fundamentals, 1979 18(1): 1115 .

55. R. C. Herndon, P. E. Hubble, J. L. Gainer, 1980 "Two Pulsators for Increasing Heat Transfer". Industrial Engineering Chemistry: Process Design and Development, 19: 405410 .

56. M. F. Edwards, W. L. Wilkinson, 1971 "Review of Potential Applications on Pulsating Flow in Pipes". Transactions of Institution of Chemical Engineers, 49: 8594 .

57. S. K. Gupta, R. D. Patel, R. C. Ackerberg, 1982 "Wall Heat /Mass Transfer in Pulsatile Flow". Chemical Engineering Science, 37(12): 17271739 .

58. L. C. Thomas, 1974 "Adaptation of the surface Renewal Approach to Momentum and Heat Transfer for Turbulent Pulsatile Flow". Journal of Heat Transfer, 348353 .

59. S. N. Kazi, 2001 "Heat Transfer to Fibre Suspensions-Studies in Fibre Characterisation and Fouling Mitigation", PhD thesis, Chemical and Materials Engineering. The University of Auckland: Auckland, New Zealand.

60. G. Parkinson, W. Price, 1984 "Getting the Most out of Cooling Water". Chemical Engineering,: 2225 .

61. D. Hasson, D. Bramson, 1985 "Effectiveness of Magnetic Water Treatment in Suppressing Calcium Carbonate Scale Deposition". Industrial Engineering Chemistry: Process Design and Development, 24: 588592 .

62. J. D. Bernardin, S. H. Chan, 1991 "Magnetics Effects on Similated Brine Properties Pertaining to Magnetic Water Treatment". Fouling and Enhancement Interactions, 28th National Heat Transfer Conference. Minneapolis, Minnesota, USA.

63. W. W. Frenier, J. Barber, 1998 1998 "Choose the Best Heat Exchanger Cleaning Method". Chemical Engineering Progress,: 3744 .

64. J. Gutzeit, 1997 “Cleaning of Process Equipment and Piping”. MTI Publication, Materials Technology Institute, St. Louis.

65. C. M. Loucks, 1958 ”Organic Acids for Cleaning Power- Plant Equipment”. Annual Meeting of ASME. New York.

66. C. Reich, 1961 1961 “Scale Removal”.: U. S. A.

67. J. F. Eberhard, R. Rosene, 1961 1961 “Removal of Scale Deposits”.: U. S. A.

68. S. N. Kazi, G. G. Duffy, Chen. X. D. And, 2010 “Mineral scale formation and mitigation on metals and a polymeric heat exchanger surface”. Applied Thermal Engineering, 30: 22362242 .

CHAPTER 6

Computational Fluid Dynamic Modeling of Horizontal Air-Ground Heat Exchangers (HAGHE) for HVAC Systems

Paolo Maria Congedo [†,*], **Caterina Lorusso** [†], **Maria Grazia De Giorgi** [†] **and Domenico Laforgia** [†]

Department of Engineering for Innovation, University of Salento, via Monteroni, Lecce I-73100, Italy

ABSTRACT

In order to satisfy the requirements of Directive 2010/31/EU for Zero Energy Buildings (ZEB), innovative solutions were investigated for building HVAC systems. Horizontal air-ground heat exchangers (HAGHE) offer a significant contribution in reducing energy consumption for ventilation, using the thermal energy stored underground, in order to pre-heat or pre-cool the ventilation air, in winter and summer, respectively. This is particularly interesting in applications for industrial, commercial and education buildings where keeping the indoor air quality under control is extremely important. Experimental measurements show that, throughout the year, the outside air temperature fluctuations are mitigated at sufficient ground depth (about 3 m) because of the high thermal inertia of the soil, the ground temperature is relatively constant and instead higher than that of the outside air in winter and lower in summer. The study aims to numerically investigate the behavior of HAGHE by varying the air flow rate and soil conductivity in unsteady conditions by using annual weather data of South-East Italy. The analysis shows that, in warm climates, the HAGHE brings a real advantage for only a few hours daily in winter, while it shows significant benefits in the summer for the cooling of ventilation air up to several temperature degrees, already by a short pipe.

KEYWORDS

geothermal; ground source heat pump; ventilation; Computational Fluid Dynamic (CFD); Zero Energy Buildings (ZEB); ground heat exchanger; high efficiency; heating and cooling of buildings; horizontal heat exchanger; typical climates

1. INTRODUCTION

The building sector represents 40% of the European Union's total energy consumption. Therefore reducing energy consumption is a priority of the climate and energy package called "20-20-20". The European Directive 31/2010 [1] upholds the concept of zero or nearly zero-energy constructions (ZEBs or n-ZEBs) which are increasingly becoming more common throughout Europe. By 31 December 2020, all new buildings must be n-ZEBs, and in particular those occupied and owned by public authorities must comply with the same criteria by 31 December 2018. New buildings must fulfill these requirements before construction starts, looking at high efficiency envelopes, the installation of renewable energy systems, heat pumps and cogeneration systems.

Many studies have analyzed several types of high energetic efficiency external walls for ZEBs; in particular in the Mediterranean climate, Baglivo *et al.* [2,3] carried out an analysis through the combination of various materials, in terms of steady thermal transmittance, periodic thermal transmittance, decrement factor, time shift, areal heat capacity, thermal admittance, surface mass, thickness.

In the literature it was already demonstrated that the substitution of traditional air heat pumps coupled with Ground Source Heat Pump (GSHPs) results in a good reduction of CO_2 emissions [4] and a good return of the investment in the case of large size applications such as commercial buildings [5]. The reduction of consumption and emissions with respect to traditional heat pump is due to the increase of the Coefficients of Performance (COP) in GSHP system, both in summer and in winter [6]. Energy studies on these features [7], together with their flexible application, noise reduction and space preservation were the engine of their success and diffusion [8,9].

Desideri *et al.* [10] showed that the operating costs necessary to heat the building with the GSHP, are lower than the ones for heating the building with a natural gas boiler. Previous studies are mainly focused on the use of horizontal water ground heat exchangers [11,12,13]. Demir *et al.* [11] carried out an experimental study to test the validity of the model; experimental and numerical simulation results, calculated using experimental water inlet temperatures, were compared. Congedo *et al.* [12] compared different configurations of horizontal water ground heat exchangers in order to define the system performance. Wu *et al.* [13] investigated both experimentally and numerically the thermal performance of slinky heat exchangers of horizontal-coupled slinky GSHP in a UK climate. Few studies concern the analysis of horizontal air ground heat exchangers. In [14] the experimental measurements and numerical simulation of an air ground source heat exchanger operating at a cold climate for a passive house ventilation system were reported.

Figure 1 shows the behavior of the outside air temperature and the ground temperature varying the depth along year 2002 by a climatic station located in the city of Lecce (Italy). Ground temperature variations are damped and phase delayed, compared to the outside air temperature ones, because of the ground thermal inertia and heat capacity (higher than air). Consequently, the ground temperature, below a certain depth, remains relatively constant throughout the year. The seasonal delay that depends on the depth and the ground composition can be as long as some weeks. Therefore, at a sufficient depth, the ground temperature is higher than that of the outside air in winter and is lower in summer. The difference in temperature between the outside air and the ground can be used as a preheating means in winter and pre-cooling in summer by a HAGHE. The idea of preheating or cooling the air by the ground is already treated by Wrobel *et al.* [15,16].

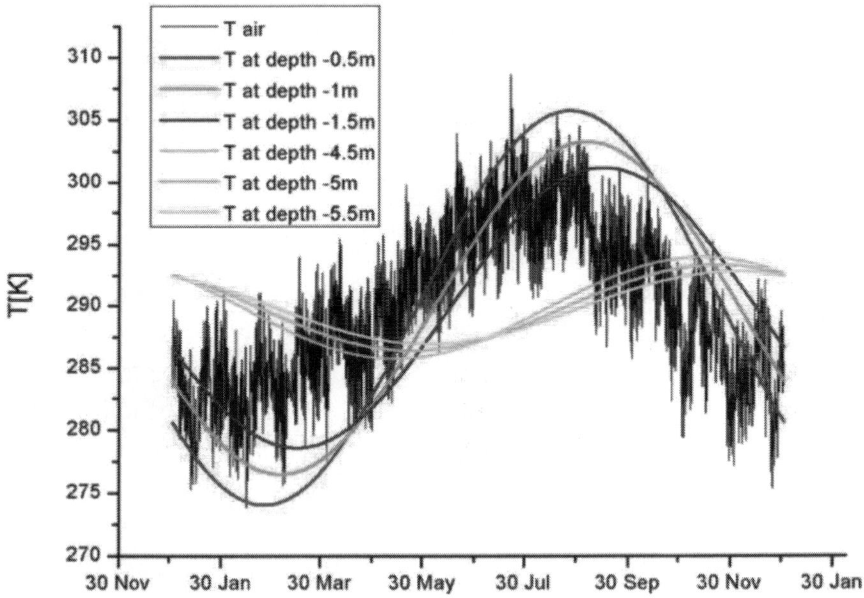

Figure 1. Ground temperature distribution for $\lambda = 2$ W/(m K), measured by a climatic station located in the city of Lecce (Italy).

HAGHE's basically consist of pipes which are buried in the ground at the depth of about 2 m, coupled with an air system which forces the outside air through the pipes and eventually mixes it with the indoor air of the building (see Figure 2). For a large number of applications it is necessary to know the influence of each parameter on the thermal performance of HAGHE as well as the pipes layout and the ground properties.

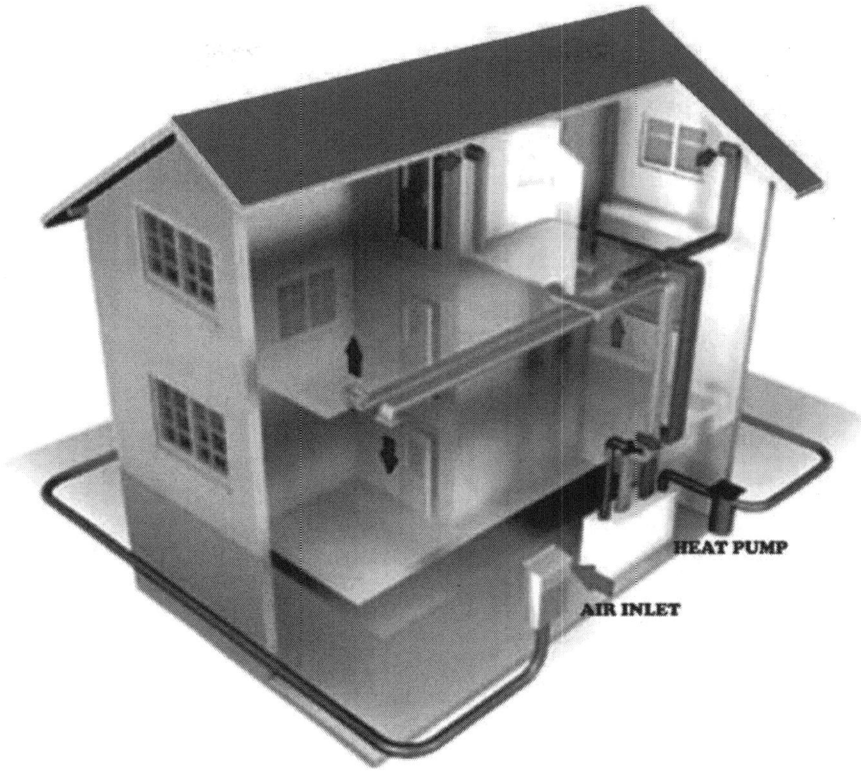

Figure 2. HAGHE integrated with mechanical ventilation or heat pump.

The present work aims to demonstrate that the geothermal heat exchanger horizontal air-to-ground (HAGHEs) can also show good performance in terms of high efficiency of the total system with a low environmental impact, operating on the ventilation air, in particular for ZEB buildings. The main cost for the installation of HAGHEs is comparable to GSHPs cost, due to drilling cost [17,18]. This study shows the thermal behavior of HAGHEs placed in south of Italy, in particular, in the city of Lecce. Some topics are discussed about the heat fluxes transferred to and from the soil and the temperature modifications of the soil around the heat exchanger, both in summer and winter.

2. CFD NUMERICAL ANALYSIS

All the simulations were performed with the CFD ANSYS FLUENT Release 14.0, which uses the control volume method [19]. Unsteady Reynolds-Averaged Navier-Stokes equations were solved with a second order implicit transient scheme in combination with the realizable k-ε model [20] with enhanced wall treatment with thermal effect option. The k-ε model is appropriate for wall-bounded and internal flows with small mean pressure gradients [21]. As a solution method pressure—velocity coupling was chosen with a Semi-Implicit Method for Pressure-Linked Equations (SIMPLE) algorithm. The CFD code main setup parameters are reported in Table 1. The numerical study was carried out in unsteady conditions with a 3D structured mesh by previous analysis [22,23]. All the geometry independent parameters were the same for all the layouts considered. The heat exchangers were simulated under different operating conditions.

Table 1. Details of CFD setting.

Element	Description
Mesh Elements	3D Structured mesh, hexahedral elements
Numbers of cells/faces/nodes	171.820/530.731/186.934
Solver	3D unsteady, pressure based
Numerical Scheme	Segregated
Linearization	Implicit
Discretization Scheme	
Pressure	PRESTO!
Pressure velocity coupling	SIMPLE
Momentum, turbulence kinetic energy, turbulence dissipation rate	2nd order upwind
Fluid	Ideal gas
Turbulence model	Realizable k-ε
Time step	1 s

2.1. Geometrical Model

A 3D structured grid was generated by means of the grid generator GAMBIT. The 3D analysis used a parallelepiped domain of 5.0 m × 2.0 m × 4.0 m (length × width × depth) with 200 mm diameter polyethylene (PE) semi-cylinder pipe, thickness 20 mm, at a depth of 2.5 m. Due to the symmetry of the investigated configuration an half of the ground-pipe heat exchanger was discretized and simulated (see Figure 3). The structured grid is composed of hexahedral cells, stretched towards the curved lateral surface of the semi-cylinder into the ground parallelepiped. The structured hexahedral mesh is typically obtained from a coarse structure (super-blocks) that divides the 2D/3D domain in parallelograms/parallelepipeds. How these super-blocks are related with each other is called topology. Depending on how the different blocks are joined with each other to cover the whole domain, are defined different topologies.Figure 4 shows the mesh of a quarter of cross section of the pipe, from the three parallelogram-shaped super-blocks, we obtain the topology "O-grid".

It was considered to discretize the area more accurate from the point of view of fluid dynamics, around the pipe, where the highest gradients are expected, using a structured mesh with hexahedral elements, towards the curved side surface of the semi-pipe. This leads to significant advantages of the numerical solution, which present a higher accuracy, and a numbering more effective. Then, a finer mesh is used (to check the accuracy of the solution) near the ground-pipe interface, which slopes down to a coarser mesh in the area far from the semi-cylinder, as shown in Figure 5. As can be clearly observed, to obtain a mesh of good quality, was strictly observed the principle of gradual change in the size of the cells; it was, in fact, imposed a "growth rate" quite low to the cells adjacent to one another, by ensuring that each cell had never more than triple in size of the previous cell.

2.2. Boundary Conditions

To model the temperature profile into the ground in function of the depth (z) and time (t), the function in Equation (1) was used [24,25], because of the interactions of the soil with sun radiation, wind and air temperature during all the year. The values of main properties of the materials used in the simulations and of the constants used in Equation (1) are reported in Table 2 and Table 3, respectively.

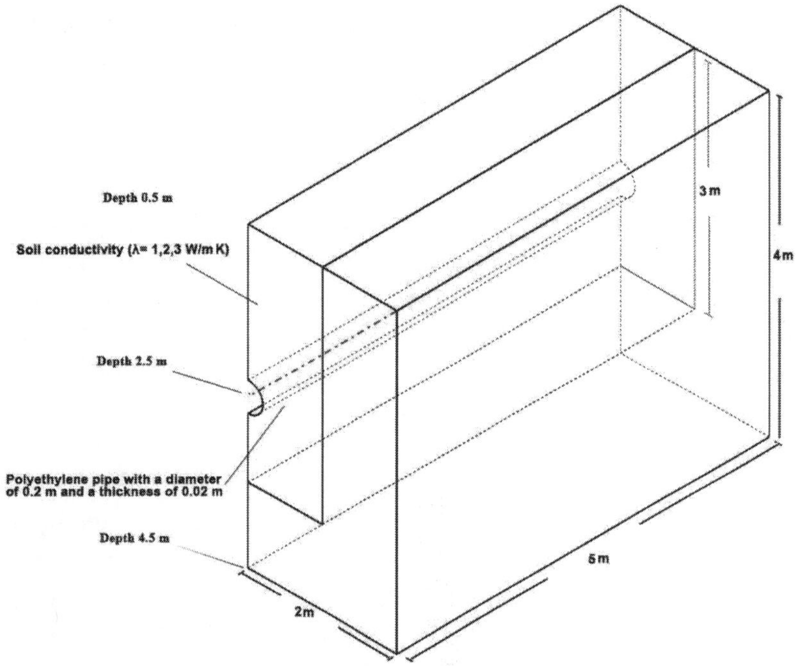

Figure 3. Simulated geometry of HAGHE due to symmetry of numerical model.

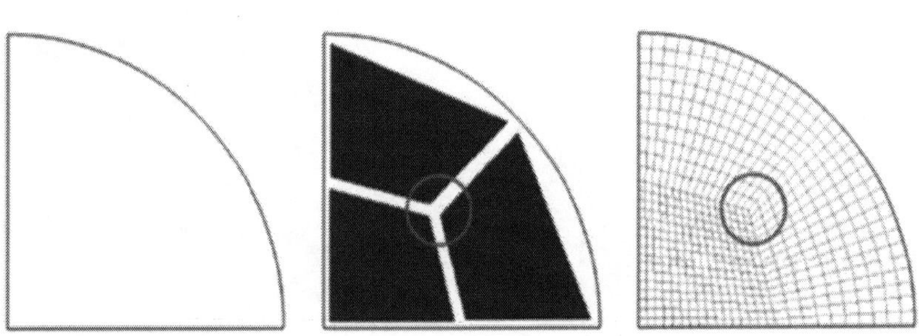

Figure 4. Mesh of a quarter of cross section of the pipe.

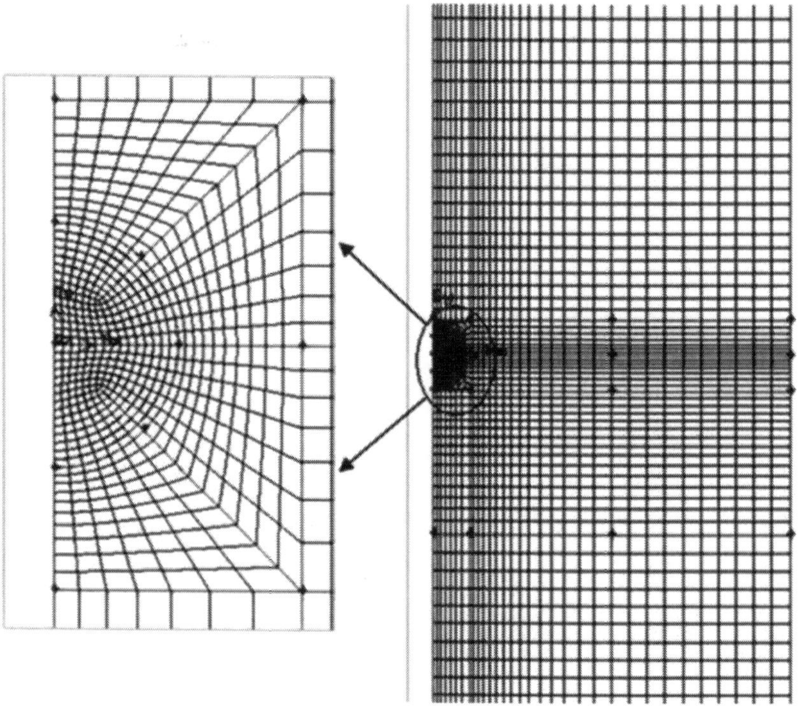

Figure 5. Details of the mesh, on left around the pipe, on right of cross section of the simulated domain.

Table 2. Main properties of the materials used in the simulations.

Variable	Pipe (PE)	Ground	Air
ρ (kg/m^3)	946	2723.23	Ideal gas
λ (W/(m·K))	0.35	1; 2; 3	0.0242
c_p (J/(kg·K))	1920	837.338	1006.43
μ(kg/s)			1.7894×10^{-5}

The values of the temperatures in the ground at different depths are reported in Figure 1 and were used, to set the boundary conditions in the ground, far from the heat exchanger and thus in undisturbed conditions, for the simulations of the operation of the ground heat exchangers, in particular at depths of 0.5 and 4.5 m. The representation by a simple harmonic of the annual cycle of the monthly mean temperature of the soil is sufficiently precise, as demonstrated by previous in-depth analysis [26,27].

In Equation (1) T_M is the mean temperature in the year of the climatic zone; a is the half difference between the maximum and minimum temperatures on soil surface in the year; τ is the considered period; t_M is the time when the maximum temperature on the ground surface occurs; D_T is the thermal diffusivity of the considered soil. The main assumptions for this equation is that all the physical properties referred to the soil are constant in space and time and thus a homogeneous soil is considered at all depth. This is a not-too-far-from-reality hypothesis because of the depth (maximum 3 m) and the climate zone (south of Italy is temperate climate zone) considered.

$$T(z,t) = T_M + a \cdot e^{-z\sqrt{\frac{\omega}{2D_T}}} \sin\left(\omega t - z\sqrt{\frac{\omega}{2D_T}} + b\right) \tag{1}$$

where:

$$\omega = 2\pi / \tau \tag{2}$$

$$b = (\tau - 4t_M)\pi / 2\tau \tag{3}$$

Table 3. Values of the constant used in Equation (1).

Constant	Value
T_M	16.73°C
a	18.74°C
Depth	2.5 m
τ	365 (1 year)
t_M	45 day

The climatic data used in order to validate the simulation were those recorded in the year 2002 by a climatic station located in the city of Lecce (Italy) at the coordinates: Latitude 40°26'16" North and Longitude 18°14'42" East. The available experimental data were air temperature and underground temperature at two depths (0.1 and 0.2 m) with a time step of an hour, the same data used in a previous work [5], where a comparison between experimental temperatures recorded in one year at two different depths in the ground with the calculated data by the CFD code was made. Figure 6 shows the hourly average temperatures at a depth of 0.2 m experimentally measured (black line) and the ones (grey line) calculated by the numerical simulation. The comparison between the two lines confirmed a good choice of the boundary condition for the numerical simulations.

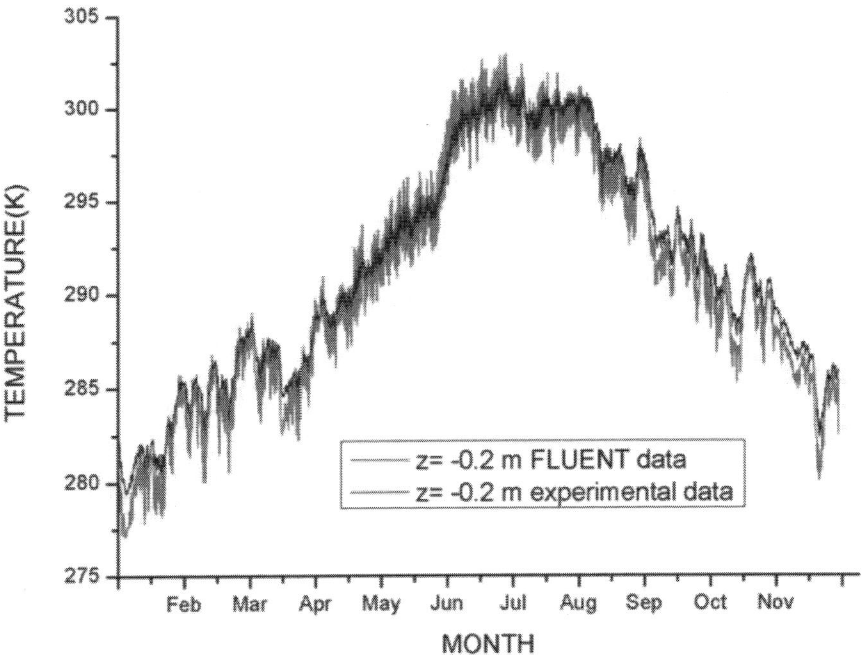

Figure 6. Comparison between the experimental data and the results from the CFD simulations [5].

Looking at Figure 3, the boundary conditions have the following setup. The upper and lower horizontal surfaces have "wall" temperatures imposed by a transient table: the soil temperatures have been imposed at depths 0.5 m and 4.5 m, respectively, varying during the year. The vertical surface close to the semi-pipe is "symmetry" and the other vertical surfaces are "adiabatic walls".

The performance of the heat exchangers were numerically investigated during the entire year, simulating the heat flux passing through a portion of soil and transmitted to the air in the polyethylene pipe. In Table 1, the main parameters of the ground properties are shown, supposed homogeneous at all depths. The investigation concerned 3 different soil types: low conductivity soil ($\lambda = 1$ W/m·K); medium conductivity soil ($\lambda = 2$ W/m·K); high conductivity soil ($\lambda = 3$ W/m·K). Weather data relating to a climatic station, located in Lecce, were generated using Meteonorm Version 5 [28].

Air flow rate was set at four different levels to investigate heat exchanger performance in different operating conditions, in particular the values were 150, 250, 350, 450 m^3/h respectively. The inlet air temperature was assumed equal to the external temperature varying hour by hour. The activation period of the heating and cooling system was, in winter, from 15 November to 31 March and each day from 8.00 am to 7.00 pm (typical for offices, factories, *etc.*), in summer, from 1 June to 31 August and each day from 8.00 am to 7.00 pm.

It is to be emphasized, that the physical problem admits, during the shutdown of the plant, when the input air flow is close to zero, the possibility of "reflux" from the air. The flow of air may, occasionally, go in the opposite direction to that of normal operation. It was, therefore, necessary to set, in the boundary condition, the back flow total temperature; it corresponds to the development of time-dependent temperature of the soil, to the level of burial of the pipe. In this way, the numerical code is able to well manage the back flow, providing the correct values during the period of shutdown of the plant.

In this work the performance of pre-heating or pre-refreshing the ventilation air using HAGHE is mainly dependent on the difference between the outdoor air temperature and the soil temperature. In South-East Italy, this difference in summer is larger than that in winter, which is the main reason why HAGHE is more effective in summer than in winter.

The analysis shows that, in warm climates, using the HAGHE has its advantages for few hours daily in winter, when the difference between the outdoor air temperature and the soil temperature is small, while it shows significant benefits in the summer for cooling of the ventilation air up to several temperature degrees, when the difference between the ground temperature and the outdoor air temperature is high (see Figure 7).

2.3. Sensitivity Analysis of Grid Spacing

In order to estimate the error due to spatial discretization a sensitivity analysis of the grid spacing was performed. Congedo *et al.*[29] show the application of the Grid Convergence Index (GCI) in industrial applications. The GCI gives a measure of the percentage the computed value is away from the value of the asymptotic numerical value. For the present work, three different grid refinement levels have been considered. Considering three grid levels, the values for the GCI indexes can be evaluated as follows:

$$GCI_{1\text{-}2} = \frac{F_s \left|\varepsilon_{1-2}\right|}{\left(r_{1-2}^{p} - 1\right)} \text{ with fine and medium grids;} \quad (4)$$

$$GCI_{2\text{-}3} = \frac{F_s \left|\varepsilon_{2-3}\right|}{\left(r_{2-3}^{p} - 1\right)} \text{ with coarse and medium grids; where} \quad (5)$$

F_s is a factor of safety and for the comparison of three grids is equal to 1.25; with:

$$h_1 < h_2 < h_3 \quad (6)$$

$$\varepsilon_{n-n+1} = \frac{f_{n+1} - f_n}{f_n} \quad (7)$$

$$r_{n-n+1} = \frac{h_{n+1}}{h_n} \tag{8}$$

$$p = \frac{\ln \dfrac{f_{n+2} - f_{n+1}}{f_{n+1} - f_n}}{\ln r} \tag{9}$$

where p is the order of convergence, h is the grid spacing (minimum value of distance between nodes of grid); f is the value of the significant parameter chosen for the comparison (the air outlet temperature has been chosen).

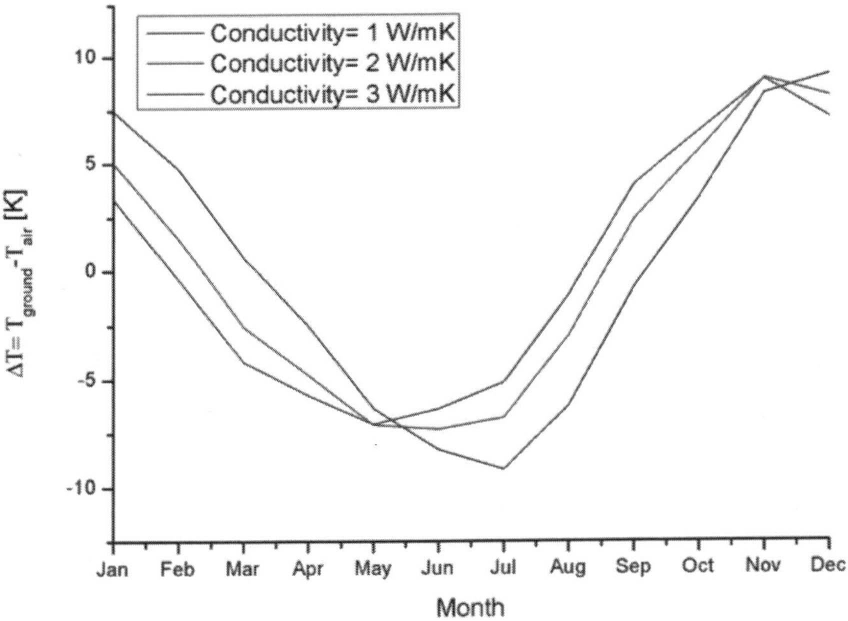

Figure 7. Difference between the ground temperature and the outdoor air temperature.

It is important that the solution of the considered grid level is in the asymptotic range; it happens if the following auditing ratio is equal to 1.

It means that increasing the number of cells does not change the value of the solution f:

$$\frac{GCI_{2-3}}{r^p \mathrm{GCI}_{1-2}} \cong 1 \tag{10}$$

Using the finest grid, the evaluated value of the solution is:

$$f = f_1 + \frac{f_1 - f_2}{r_{1-2}^p - 1} \tag{11}$$

For the present work, the air outlet temperature has been chosen as the comparison parameter. In Figure 8 such a temperature is plotted as function of time for different grid spaces (h_1= 1.65 mm; h_2 = 3.30 mm; h_3 = 6.60 mm). In detail, f_1 curve represents the temperature calculated with the finest grid, f_2 curve with the medium grid, and f_3 with coarsest one. The curves show a little divergence with increasing time. In Figure 9 the grid convergence indexes, the order of convergence and the auditing ratio are reported. It is important to verify that the auditing ratio is really close to 1.

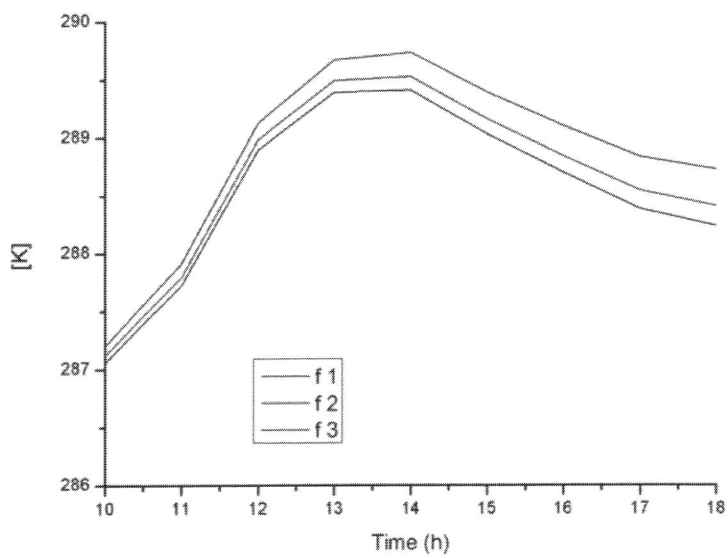

Figure 8. Air outlet temperature is plotted as function of time for different grid spaces.

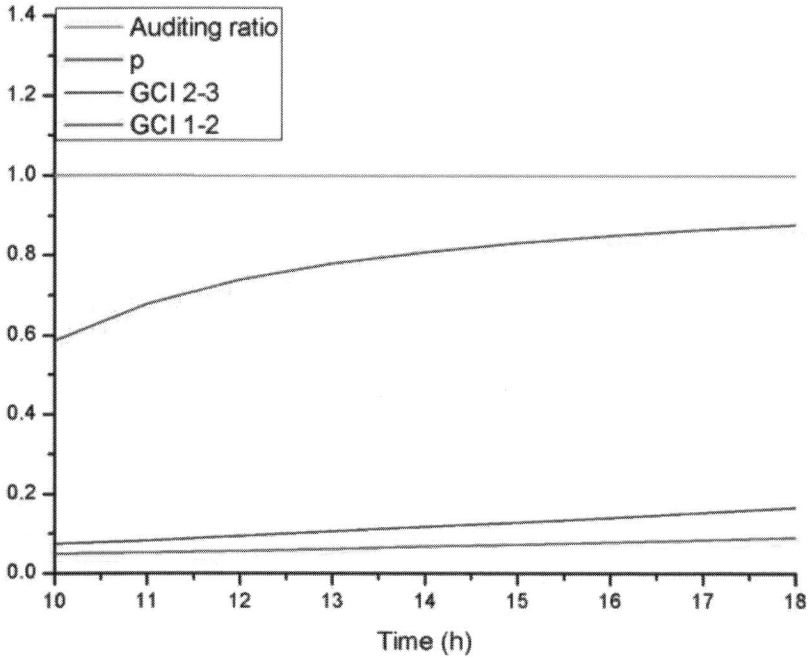

Figure 9. Grid convergence indexes, auditing ratio and order of convergence *vs.* time (grid spacing analysis).

2.4. Sensitivity Analysis of Time Step

An analogous approach has been taken for the sensitivity analysis of the time step. Looking at Figure 10 and Figure 11, again the air outlet temperature has been chosen as the comparison parameter. In detail, f_1 curve represents the temperature calculated with the smallest time step (1 s), f_2 curve with the medium time step (2 s) and f_3 with the greatest time step (4 s). Such curves are overlap completely and the auditing ratio trend in Figure 11 confirms that the solution is in asymptotic range.

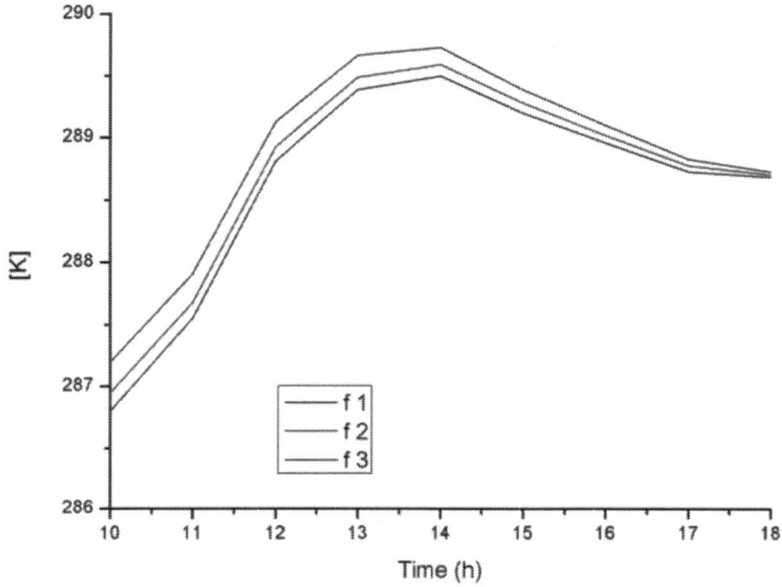

Figure 10. Air outlet temperature is plotted as function of time for different time steps.

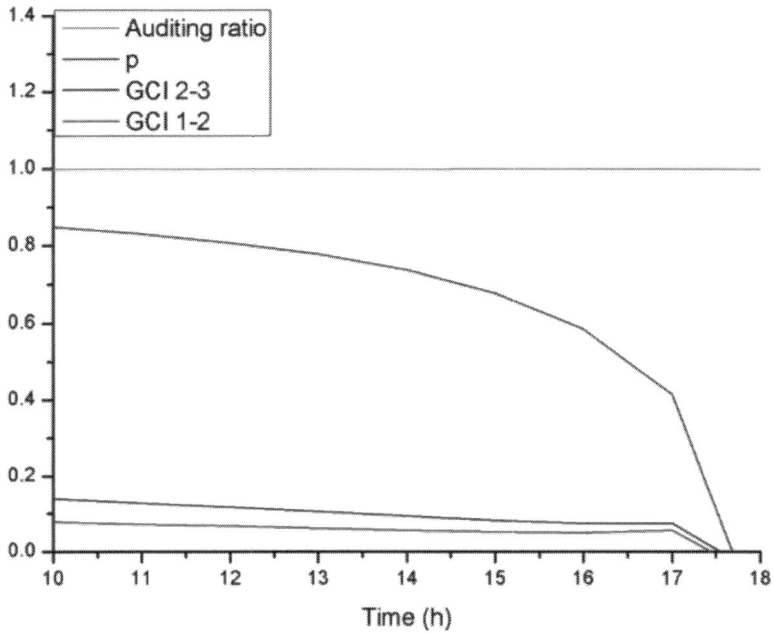

Figure 11. Grid convergence indexes, auditing ratio and order of convergence *vs.* time (time step analysis).

3. RESULTS

The CFD simulations showed the HAGHE's behavior (with a length of the pipe, as said, of 5 m) in different operating conditions. As regards the analysis relating to the data of Lecce, the results are represented into four figures: the first three Figure 12, Figure 13 andFigure 14 show the performances of the air-ground heat exchanger in terms of ground and inlet-outlet air temperatures and air-ground thermal power, by varying air flow rate (150 to 450 m^3/h), keeping constant the value of soil conductivity and assuming the temperature of the ground, T_{ground}, with the plant shut down. Figure 15 shows, instead, the HAGHE performance by varying the soil conductivity (1, 2, 3 W/m K), keeping constant air flow rate at 350 m^3/h.

The performance results were represented for the weeks in which climatic conditions are more "strict"; for winter, the week with the minimum peak temperature of the year is from 11 to 17 February, in summer, the week with the maximum peak temperature of the year is from 19 to 25 July (year 2002). During both weeks in winter and summer, the outlet air temperature shows, for each daily shutdown of the geothermal plant, an asymptotic behavior at the ground temperature, this henceforth will be called base-line. The value of the base-line depends on the thermal conductivity of the ground; in fact, the base-line tends to overlap the temperature of the ground faster, by increasing thermal conductivity (see Figure 12a and Figure 14a).

The principle that the air is in thermal equilibrium with the ground, during the shutdown of the plant is, certainly, plausible and well managed by the simulation in the CFD code, according to the settings on the back flow. The base line has a correlation, in the diagrams of heat flux, with the value of zero flux. The heat flux is related to the entire pipe and not to semi-pipe. The heat flux is assumed negative, when the air circulating in the duct transfers heat to the ground.

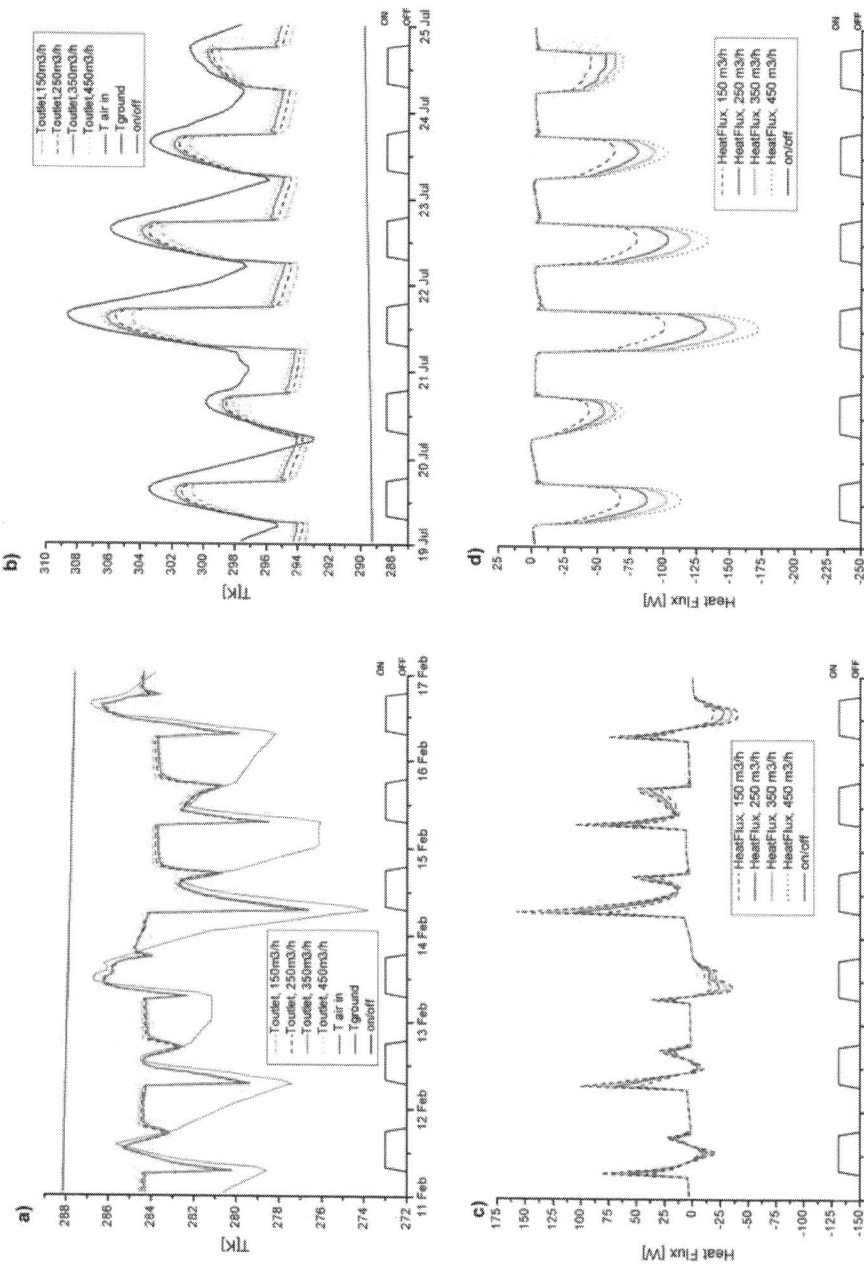

Figure 12. Ground and inlet-outlet air temperatures with $\lambda = 1$ W/m K, air-ground thermal power varying the air flow rate (range 150–450 m^3 per hour), for typical winter (**a,c**) and summer weeks (**b,d**).

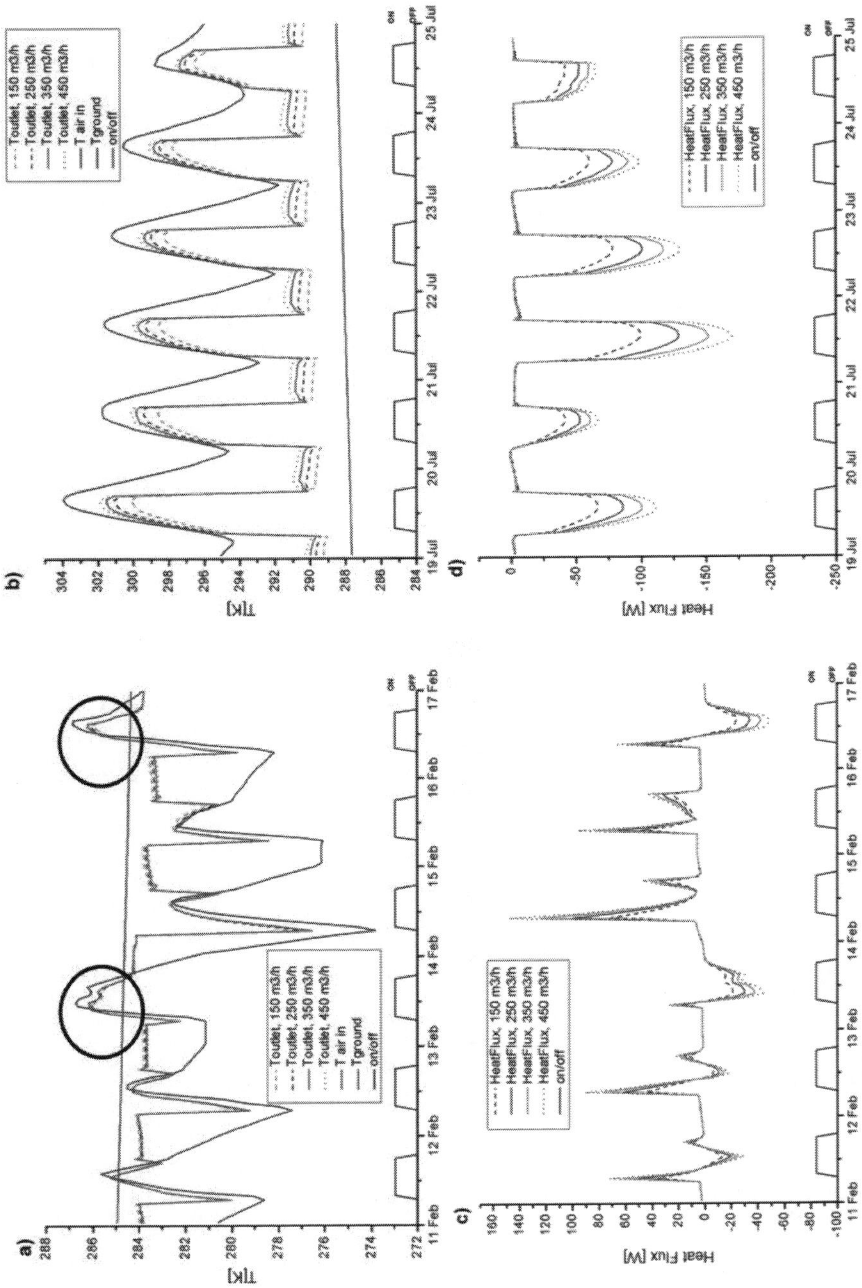

Figure 13. Ground and inlet-outlet air temperatures with λ = 2 W/m K, air-ground thermal power varying the air flow rate (range 150–450 m^3 per hour), for typical winter (**a**,**c**) and summer weeks (**b**,**d**).

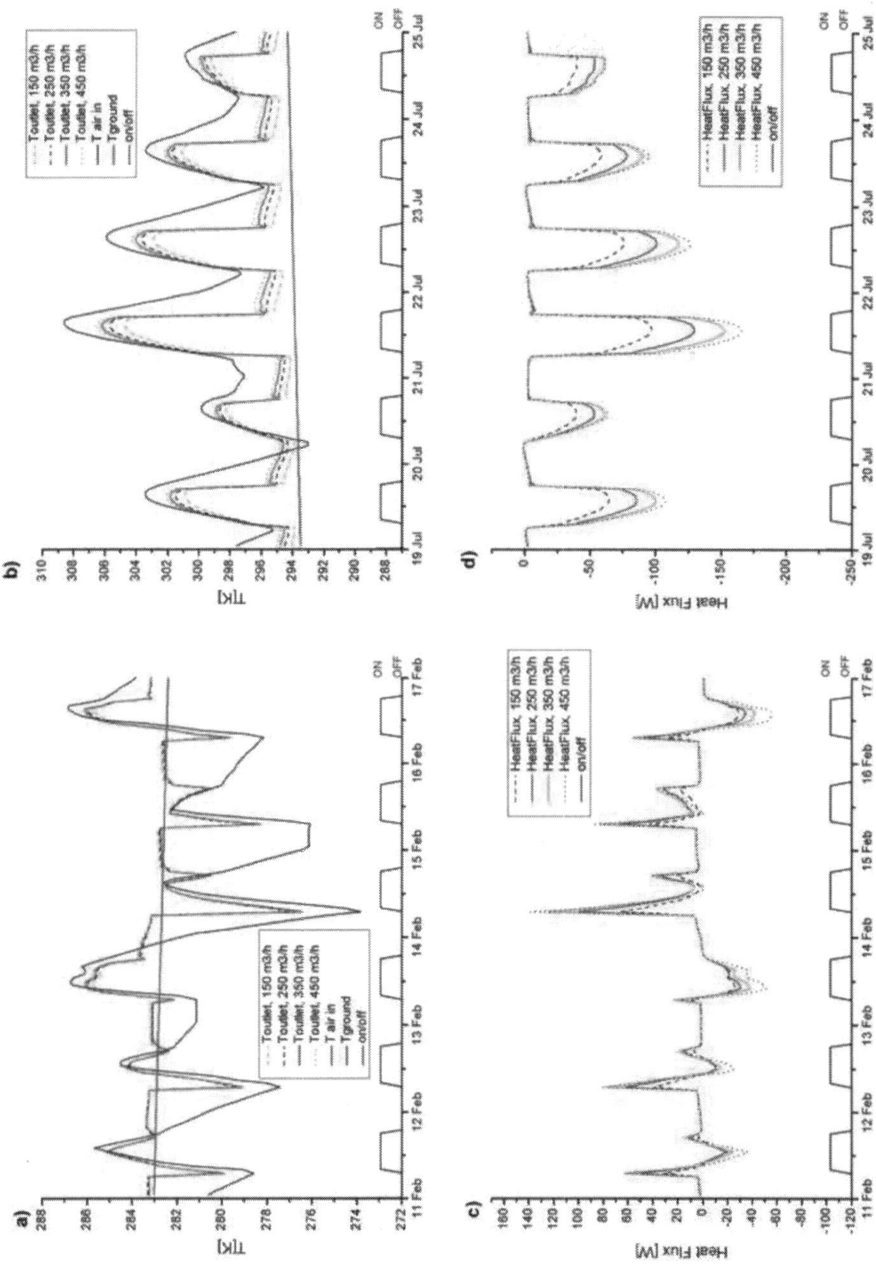

Figure 14. Ground and inlet-outlet air temperatures with $\lambda = 3$ W/m K, air-ground thermal power varying the air flow rate (range 150–450 m^3 per hour), for typical winter (**a**,**c**) and summer weeks (**b**,**d**).

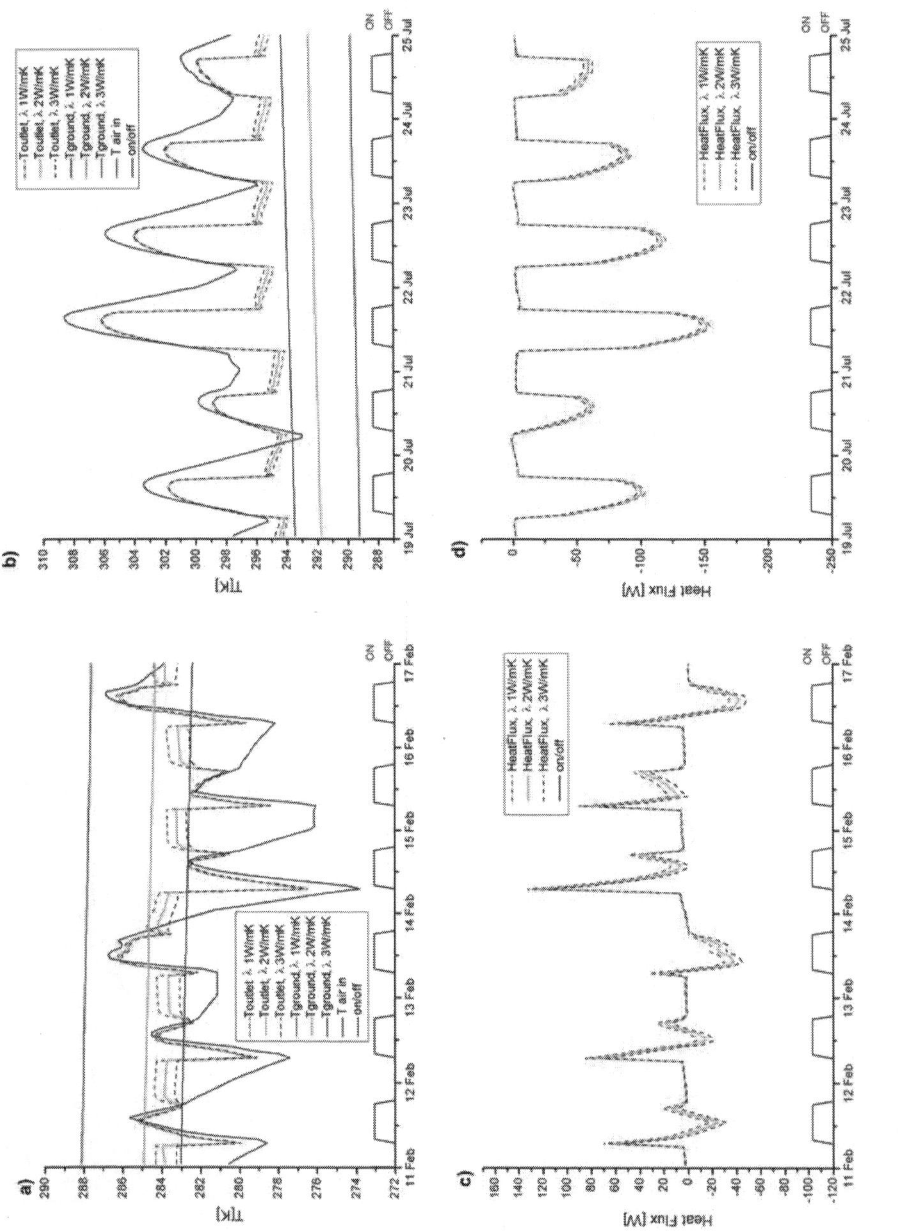

Figure 15. Inlet-outlet air temperatures, air-ground thermal power varying the soil conductivity λ = 1–3 W/m K, fixed air flow rate 350 m^3/h, for typical winter (**a**,**c**) and summer weeks (**b**,**d**).

The trends of outlet air temperatures and the corresponding heat fluxes show an unsuitable behavior of the geothermal plant, during the winter. Although 11–17 February is a cold week, the results show the presence of negative peaks in heat transfer, in almost every ON-block. This means that the plant is not able to handle the heating of inlet air, during the whole daily ON-block (the classic office hours). It can happen that, often around 12:00 am, the inlet air temperature "overtakes" the ground temperature, making the reversal of flow (see Figure 13a).

Instead, the behavior of the HAGHE in summer is significantly positive; with the system switched on, the heat fluxes are always negative in order to pre-cooling the inlet air, depending on the air flow rate. Figure 12b and Figure 14b, show a temperature decrease of 0°–4°, according to the air flow rate, with a pipe only 5 m long.

In Figure 12b, observing the peak relative to the 21 July, can be seen that the air is undisturbed at a temperature of 309 K, while the out of the plant goes from 305 K, for a flow rate of 150 m^3/h, to 307 K, for a flow rate of 450 m^3/h. Clearly, a lower volumetric flow rate input, which means lower air velocity, confer the possibility of a more efficient heat exchange with the ground, for a long probe 5 m; the results are a cooler temperature and a better conditioning, for a flow rate of 150 m^3/h.

Finally, Figure 15 shows the HAGHE performance by varying the soil conductivity (1, 2, 3 W/m K), keeping constant air flow rate at 350 m^3/h. The first evident difference between winter and summer is that in summer the heat flux is always in the same direction instead in winter the heat flux inversion is possible. Apparently there is a contradiction between Figure 15a–c because the heat flux grows with the ground thermal conductivity but the outlet air temperature is highest with the lowest value of it. The reason is that ground with lower thermal conductivity is more isolated from the external weather conditions, in winter the ground temperature is higher with lower ground thermal conductivity.

A possible realistic solution to improve the efficiency of HAGHE buried in soil with low thermal conductivity is filling the trench with high thermal conductive materials. Figure 16a,b show the pressure drops along pipe 5 m long, for winter and summer, respectively.

Concluding the analysis underlines that, in warm climate, during the winter, the HAGHE can be used for a few hours daily and must have a bypass for the external air when the air temperature is higher than one of ground. Instead, during the summer, the HAGHE has huge advantages for cooling the ventilation air. A possible problem should be the relative humidity of air at the end of pipe, therefore it is important to evaluate for each case, if there is a need to keep under control the humidity by a heat pump. In particular, in the historical buildings with moisture dynamics implications [30,31], good ventilation using dry air seems to be necessary.

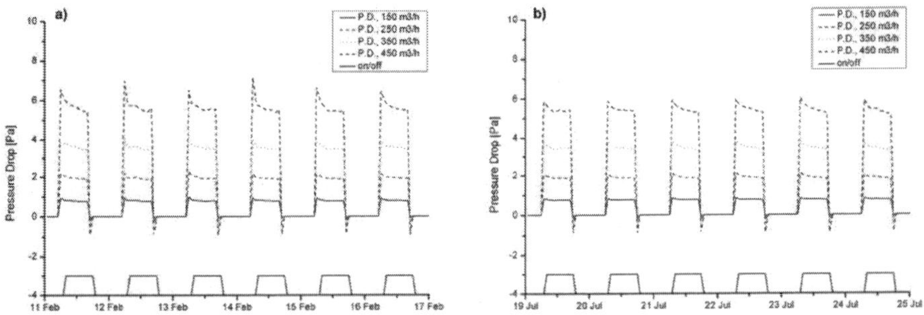

Figure 16. Pressure drops of air *vs.* time varying the air mass flow rate for typical: (**a**) winter week and (**b**) summer week.

4. CONCLUSIONS

This work aims to investigate the performance of HAGHE for HVAC systems. By using the thermal energy stored in the ground, it is possible to pre-heat or pre-cool the ventilation air. HAGHE is a simple superficial system, because the pipes are not buried deep into the ground with air like vector fluid (see Figure 17). The horizontal geometry of pipes guarantees good heat transfer between air and ground and with a useful stability of air temperature. The horizontal geometry outperforms the vertical configuration, has not the problems related to deep wells, like cost and risk to meet the aquifer.

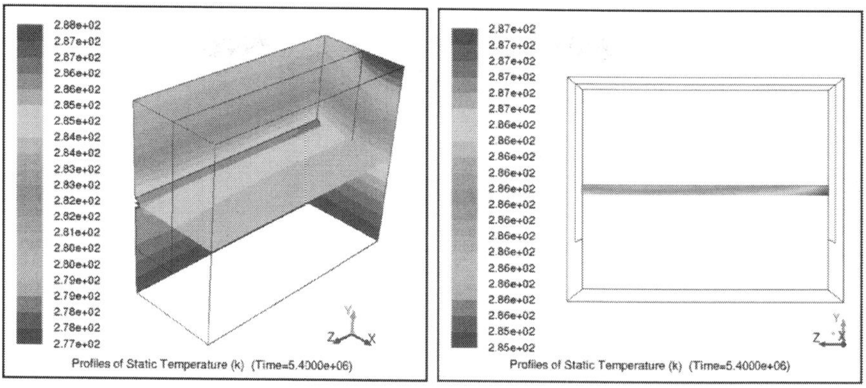

Figure 17. Temperature distribution on the horizontal and vertical planes, air flow rate equal to 450 m^3/h, with $\lambda = 3$ W/(m K), 20 February, at 12 a.m.

It is interesting to underline that such a condition offers a different strategic possibility, which is the application of HAGHE not for ventilation air but with an air-source heat pump, using the air flux from the HAGHE to keep the behavior of the heat pump stable, reducing the air temperature fluctuations.

The analysis of the results underlines that the HAGHE can be used for a few hours daily with a bypass for the external air when the air temperature is higher than the ground temperature. A smaller volumetric flow rate input, which means lower air velocity, leads to a more efficient heat exchange with the ground, for a 5 m long probe; this results in a cooler temperature and a better conditioning, for a flow rate of 150 m^3/h. A comparison at equal flow of air, in function of the conductivity of the soil, shows a slight loss of performance, with the increase of the conductivity value. It is recommended that the heat exchanger be in ground insulation, and consequently less sensitive to thermal changes external; the solution could be a simple conductive trench, limited in size, which surrounds the heat exchanger.

The simulations show significant benefits of HAGH during the summer. The plant is able to process the inlet air during the whole day. The decrease of the outlet air temperature is between 2° and 4°, with a pipe only 5 m long. This offers a good possibility to use the HAGHE for a thermal pre-treatment of ventilation air. Of course the system shows good performance in terms of efficiency and environmental impact if coupled in buildings meeting the n-ZEB or ZEB standards.

Further numerical investigations will be focused on experiments that will be also performed to collect data for a better understanding of the heat transfer phenomenon with the ground, in real operating conditions for a long time.

ACKNOWLEDGMENTS

The authors thank Simone Occhilupo for his powerful suggestions.

AUTHOR CONTRIBUTIONS

Paolo Maria Congedo planned the work, Paolo Maria Congedo and Maria Grazia De Giorgi carried out the CFD analysis, Caterina Lorusso did the analysis of results and wrote the paper, Domenico Laforgia did the revision of the paper.

CONFLICTS OF INTEREST

The authors declare no conflict of interest.

REFERENCES

1. Directive 2010/31/EU of the European Parliament and of the Council of 19 May 2010 on the Energy Performance of Buildings; European Union: Brussels, Belgium; 18; June; 2010.

2. Baglivo, C.; Congedo, P.M.; Fazio, A.; Laforgia, D. Multi-objective optimization analysis for high efficiency external walls of zero energy buildings (ZEB) in the Mediterranean climate. *Energy Build.* **2014**, *84*, 483–492.

3. Baglivo, C.; Congedo, P.M.; Fazio, A. Multi-criteria optimization analysis of external walls according to ITACA protocol for zero energy buildings in the mediterranean climate. *Build. Environ.* **2014**, *82*, 467–480.

4. Genchi, Y.; Kikegawa, Y.; Inaba, A. CO_2 payback-time assessment of a regional-scale heating and cooling system using a ground source heat-pump in a high energy-consumption area in Tokyo. *Appl. Energy* **2002**, *71*, 147–160.

5. Starace, G.; Congedo, P.M.; Colangelo, M.G. Horizontal heat exchangers for GSHPs efficiency and cost investigation for three different applications. In Proceedings of the ECOS 2005 conference, Trondheim, Norway, 20–23 June 2005.

6. Lazzarin, R. Ground as a possible heat pump source. *Geothermische Energie* **2001**, *24–31*, 32–33.

7. Akpinar, E.K.; Hepbasli, A. A comparative study on exergetic assessment of two ground-source (geothermal) heat pump systems for residential applications. *Build. Environ.* **2007**, *42*, 2004–2013.

8. Sanner, B.; Karytsas, C.; Mendrinos, D.; Rybach, L. Current status of ground source heat pumps and underground thermal energy storage in Europe. *Geothermics* **2003**, *32*, 579–588.

9. Kavanaugh, S.P.; Rafferty, K. *Ground Source Heat Pumps—Design of Geothermal Systems for Commercial and Institutional Buildings*; American Society of Heating Refrigerating and Air-Conditioning Engineers (ASHRAE): Atlanta, GA, USA, 1997.

10. Desideri, U.; Sorbi, N.; Arcioni, L.; Leonardi, D. Feasibility study and numerical simulation of a ground source heat pump plant, applied to a residential building. *Appl. Therm. Eng.* **2011**, *31*, 3500–3511.

11. Demir, H.; Koyun, A.; Temir, G. Heat transfer of horizontal parallel pipe ground heat exchanger and experimental verification.*Appl. Therm. Eng.* **2009**, *29*, 224–233.

12. Congedo, P.M.; Colangelo, G.; Starace, G. CFD simulations of horizontal ground heat exchangers: A comparison among different configurations. *Appl. Therm. Eng.* **2012**, *33–34*, 24–32.

13. Wu, Y.; Gan, G.; Verhoef, A.; Vidale, P.L.; Gonzalez, R.G. Experimental measurement and numerical simulation of horizontal-coupled slinky ground source heat exchangers. *Appl. Therm. Eng.* **2010**, *30*, 2574–2583.

14. Flaga-Maryanczyk, A.; Schnotale, J.; Radon, J.; Was, K. Experimental measurements and CFD simulation of a ground source heat exchanger operating at a cold climate for a passive house ventilation system. *Energy Build.* **2014**, *68*, 562–570.

15. Wrobel, J.; Ma, X.; Schmitz, G.; Grabe, J. A desiccant assisted air conditioning system with use of geothermal energy. In Proceedings of the of World Geothermal Congress, Bali, Indonesia, 25–29 April 2010.

16. Wrobel, J.; Schmitz, G. Geothermal—And solar assisted air conditioning system. In Proceedings of the 14th International Refrigeration and Air Conditioning Conference at Purdue University, West Lafayette, IN, USA, 16–19 July 2012; pp. 1–10.

17. Bloomquist, R.G. The economics of geothermal heat pump systems for commercial and institutional buildings. In Proceedings of the Conference on Geothermal Energy in Underground Mines, Ustron, Poland, 21–23 November 2001.

18. Badescu, V. Economic aspects of using ground thermal energy for passive house heating. *Renew. Energy* **2007**, *32*, 895–903.

19. ANSYS Inc. *ANSYS FLUENT Release 14.0 User Manual*; ANSYS Inc.: Canonsburg, PA, USA, 2011.

20. Shih, T.H.; Liou, W.W.; Shabbir, A.; Yang, Z.; Zhu, J. A new k-ε eddy viscosity model for high Reynolds number turbulent flows. *Comput. Fluids* **1995**, *24*, 227–238.

21. Periasamy, M.; Sivakumar, D.B.; Kumar, T.S. Dampening flow induced vibration due to branching of duct at elbow. *Int. J. Comput. Eng. Res.* **2012**, *2*, 1001–1004.

22. Fu, Y.Z.; Du, X.H.; Zhang, X. Design and simulation analysis of pile foundation ground source heat exchanger system. *Appl. Mech. Mater.* **2014**, *448–453*, 2815–2818.

23. Lenhard, R.; Malcho, M. Numerical simulation device for the transport of geothermal heat with forced circulation of media.*Math. Comput. Model.* **2013**, *57*, 111–125.

24. Cavazza, L. *Fisica del Terreno Agrario*; Unione Tipografico-Editrice Torinese (UTET): Florence, Italy, 1981; pp. 203–295.

25. Kirkham, D.; Powers, W.L. *Advanced Soil Physics. J.*; Wiley-Interscience: New York, NY, USA, 1972.

26. Penrod, E.B. *Variation of Soil Temperature at Lexington, Kentucky, from 1952–1956*; Engineering Experiment Station, College of Engineering, University of Kentucky: Lexington, KY, USA, 1960.

27. Carson, J.E. Analysis of Soil and Air Temperature by Fourier Techniques. *J. Geophys. Res.* **1963**, *68*, 2217–2232.

28. METEONORM. Available online: http://www.meteonorm. com (accessed on 15 November 2008).

29. Congedo, P.M.; Ficarella, A.; Laforgia, D. Fluid-dynamic analysis and optimization of the quenching process for hardening of change-speed gears using DOE-ANOVA method. *J. Heat Transf.* **2004**, *126*, 365–375.

30. D'Agostino, D.; Congedo, P.M. CFD modeling and moisture dynamics implications of ventilation scenarios in historical buildings. *Build. Environ.* **2014**, *79*, 181–193.

31. D'Agostino, D.; Congedo, P.M.; Cataldo, R. Computational fluid dynamics (CFD) modeling of microclimate for salts crystallization control and artworks conservation. *J. Cult. Herit.* **2014**, *15*, 448–457.

CHAPTER 7

Thermal Efficiency Comparison of Borehole Heat Exchangers with Different Drillhole Diameters

Jin Luo [1,*], Joachim Rohn [1], Manfred Bayer [2] and Anna Priess [1]

[1]GeoZentrum Nordbayern, Friedrich Alexander University Erlangen Nuremberg, Bayern 91054, Germany

[2]TÜV Rheinland, LGA Bautechnik GmbH, Tillystr. 2, 90431 Nuremberg, Germany

ABSTRACT

Thermal efficiency of borehole heat exchangers (BHE) is of crucial importance for the design and optimization of ground source heat pump (GSHP) system. This paper investigates thermal efficiency of a BHE with three drillhole diameters: 121 mm, 165 mm and 180 mm. The BHE was installed in a GSHP system of an office building located in Nuremberg, Germany. Thermal properties and hydraulic properties of the ground where the BHE was installed have been measured by thermal response tests as well as pumping tests. Furthermore, the evaluation of thermal performance is made possible by monitoring operation of the

GSHP system. Using the recorded data, thermal exchange rates have been calculated and compared in a daily period as well as a seasonal period. The daily statistics indicate that the thermal exchange rate of the BHE increases with larger drillhole diameter. For the seasonal cooling performance, the amount of thermal exchange of BHE with 165 mm and 180 mm diameters was found to be 3.2% and 7.1% larger than that of the BHE with 121 mm diameter, respectively. These findings provide helpful suggestions for the design of future GSHP systems to achieve higher energy-efficiency.

KEYWORDS

ground source heat pump system; borehole heat exchanger; drillhole diameter; thermal efficiency

1. INTRODUCTION

Recently, the increasing usage of fossil fuels has been inducing global warming as well as air pollution. Air conditioning of buildings accounts for a large part of the worldwide fossil energy consumption. In order to solve such pending problems, several new energy-efficient solutions have been developed in the last decades. Among all the technologies, ground source heat pump (GSHP) systems have been spotlighted as an efficient energy system due to their great energy savings and CO_2 emissions reduction potential. The thermal performance of GSHP system depends heavily on the heat transfer between a borehole heat exchanger (BHE) and its surrounding soil/rock. To ensure high efficiency of GSHP systems, thermal efficiency of the BHE is of essential importance [1,2,3].

Thermal performance of a GSHP system is commonly estimated by *in-situ* monitoring or by numerical modeling of the BHE. In the last decades, several studies [4,5,6,7,8,9] have been performed to assess the thermal performance of BHEs. These studies have discussed various factors affecting thermal performance such as geometric configuration of the BHE, ground conditions, *etc*. Pahud and Matthey [10] compared the thermal performance of a double U-tube BHE with different back fillings. Thermal performance was found to be enhanced with increasing thermal conductivity of the back fillings. Similarly, Borinaga-Treviño *et al.* [11] reported a study concerning the borehole thermal response and thermal resistance of BHE with four different grouting materials. The results indicated that the BHE with a cement-bentonite-graphite based grouting performs better than that of the BHE with cement-based grouting materials. The laboratory-measured grout thermal conductivity was found not to be suitable for borehole resistance estimation due to the increase of contact resistance between pipe and grout during the shrinkage of the cement based mix. Lee *et al.* [12] investigated the effective thermal conductivity in closed-loop vertical ground heat exchangers. Thermal efficiency was compared for six boreholes constructed with different grouting materials, different configurations as well as different additives. The testing results revealed that cement grouting has a higher effective thermal conductivity than that of bentonite grouting, and graphite performs better than silica. Besides, less thermal interference was found between the inlet and outlet pipe of a new 3-pipe type heat exchanger as compared to the conventional U-tube heat exchanger. Desmedt *et al.* [13] demonstrated an experimental study of thermal performance of borehole heat exchangers with different grouting materials. Thermal conductivity and borehole resistance were both measured *in-situ* by TRT tests. The borehole resistance for a standard coaxial tube with a cement-bentonite grouting was measured to be 0.344 K·m/W, whereas this value for a double U-tube was estimated to be 0.162 K·m/W which accounts for 48% of the coaxial tube.

Besides the grouting material, BHE configurations are also important for thermal performance. Zeng *et al.* [14] proposed an analytical solution considering the impacts of fluid axial convective heat transfer on borehole configurations. Different borehole configurations and flow circuit arrangements in regard to their thermal resistance have been assessed using this method. Results showed that the double U-tube performs better than a single U-tube. However, an opposite conclusion was obtained by Shu *et al.* [15] during the evaluation of thermal output for single and double U-tube with taken into account the economic performance. Therefore, both thermal and economic performance needs to be considered for an investment of a BHE. Choi *et al.* [16] investigated the effects of ground water flow on BHE arrays. Three types of BHE arrays have been examined using a two-dimensional numerical model. The results indicated that thermal performance of L-type and single line arrays are noticeably influenced by groundwater flow direction as well as flow rate.

The groundwater flow has also been reported to be an important factor in the thermal performance of BHEs. Wang *et al.* [17] demonstrated an experimental setup for analysis of the thermal performance of a BHE with consideration of groundwater flow. A model was developed to estimate the groundwater flow characteristics. The results showed that the groundwater flow has obvious impacts on the ground temperature profile in aquifers, which improves the thermal performance of the BHE. Zanchini *et al.* [18] examined the long-term performance of large borehole heat exchanger fields with unbalanced loads and groundwater flow. Peclet numbers in the range of 0–0.8 were considered in this study. The results show that the hourly peak loads was not affected by groundwater flow. However, the groundwater flow has an obviously influence on the long-term performance. On the other hand, Lazzari *et al.* [19] investigated the long-term performance of BHE fields with a negligible groundwater movement. The impacts of six time periodic heat loads on different field geometries were studied numerically. The findings indicated that no compensation between winter and summer loads is necessary for a single BHE. For a line of BHEs and for two staggered lines of staggered BHEs, at least a partial compensation of winter heating and summer cooling is needed.

To analyze in depth the thermal performance of a BHE, the short time response of a BHE as well as the thermal interaction between adjacent BHEs also need to be assessed. Lamarche and Beauchmp [20] developed an analytical model for the short time transient response of a BHE. Less than one hour heat transfer of BHE can be simulated using this model. The results from this analytical model follow very well the results from numerical models that take into account the thermal capacity of the borehole. Bernier *et al.* [21] studied the thermal interaction among boreholes in a geo-exchange system with vertical ground heat exchangers. The results indicated that long-term temperature changes have a significant impact on borehole field sizing and energy simulation results. In addition, the concept of temperature penalty which represents an effective increase of the undisturbed ground temperature was used for evaluation of borehole thermal interactions. A correlation was proposed to predict the temperature penalty and it provided results within an uncertainty of ±10% for most operation conditions.

However, the aforementioned studies were limited to small scales such as laboratory measurements and site tests. Some of them examined the thermal exchange in the BHE only based on numerical simulation. In this work, a more thorough study of thermal efficiency of BHE is conducted under practical GSHP system operation conditions. An analysis of the thermal efficiency of the BHE has been conducted on a GSHP system installed in an office building located in Nuremberg, Germany. The BHE with three drillhole diameters were installed in the GSHP system. The thermal performance is analyzed using the monitoring data accumulated under operation of the GSHP system. Thermal exchange between the BHE and its surrounding soil/rock is measured and analyzed. The remaining sections of this paper are organized as follows: Section 2 presents the ground properties measurement and the system configuration. Section 3 describes the measured data and analyzes the thermal performance of BHE under different diametric conditions. Finally, the article concludes with a summary of the results of this study in Section 4.

2. EXPERIMENTAL MEASUREMENTS

2.1. Ground Properties Measurement

2.1.1. Thermal Properties

Thermal properties of the ground where the BHE was installed have been measured by performing thermal response tests (TRT). For a TRT test, a defined heat load is input in the circulating heat carrier fluid and the resulting temperature changes are measured [22].Figure 1 shows the recording of the TRT test performed for the drillhole of 121 mm diameter [23]. It is observed that the fluid temperature keeps increasing with testing time. Heat input remains almost constant during the whole testing period, except for a few minutes in the beginning of the tests. At the same time, the ambient temperature was observed to change abruptly, ranging from 3 °C to 30 °C during the 7-day testing period. These findings imply that the abrupt change in ambient temperature is an important factor affecting the testing results.

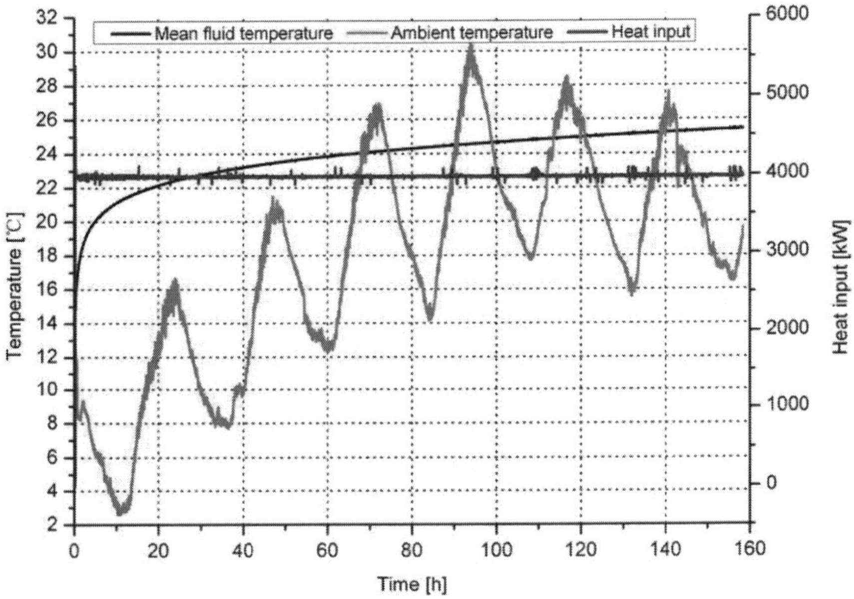

Figure 1. Recording of the TRT test performed in the drillhole of 121 mm diameter. The TRT tests have been performed separately at the three ensembles of the installed BHEs.

The TRT measurements were interpreted by following Hellström's method [24]. In this method, the heat transfer of a BHE in the ground is described by two parameters: effective thermal conductivity which describes the mean thermal conductivity of the surroundings for borehole, λ_{eff}; and borehole resistance which describes heat transfer between pipes and borehole wall, R_b. A steady-state heat transfer calculations gives a heat flow Q, between circulating fluid and borehole wall. Then, the heat transfer in a BHE can be derived as follows:

$$T_f = \frac{Q}{4\pi\lambda_{eff} H}\ln(t) + \frac{Q}{H}\left\{\frac{1}{4\pi\lambda_{eff}}\left[\ln\left(\frac{4a}{r_0^2}\right) - \gamma\right] + R_b\right\} + T_s \tag{1}$$

where *m* is the inclination of the curve of temperature versus logarithmic time, to calculate effective thermal conductivity; λ_{eff}, the formula has to be transformed:

$$\lambda_{eff} = \frac{q}{4\pi Hm} \tag{2}$$

In the first few hours, the temperature development is mainly controlled by the borehole filling and not by the surrounding soil or rock. Therefore, the minimum duration criterion should be noted. The minimum time for valid evaluation of the test is recommended as:

$$t > \frac{5r^2}{\alpha} \tag{3}$$

The measured thermal conductivities for the three different drillhole diameters match nicely, which means average thermal conductivity along the borehole is around 2.5–2.6 W/m/K, as summarized in Table 1. Besides, drillhole diameter has an effect on the R_b, with bigger diameters yielding higher resistance. Similar findings were obtained by Smith and Perry [7] for tests conducted on BHEs using different grouts and varying pipe and drillhole diameters.

Table 1. Results obtained from the performed three TRT tests.

Drillhole diameter	λ (W/m/K)	R_b (m*K/W)
121 mm	2.5	0.093
165 mm	2.6	0.105
180 mm	2.6	0.110

2.1.2. Hydraulic Properties

To measure hydraulic properties, pumping tests have been performed. In order to perform the pumping tests, two investigation wells have been drilled with a depth of 80 m. One of the wells, the upstream, was named as groundwater measurement point one (GWM 1). The other one which located at the downstream was named as groundwater measurement point two (GWM 2).

According to the material types, the pumping tests have been performed in three different geological layers (Figure 2). Packers were used at each boundary between different layers to separate the tests. Thiem's method was used for interpretation of the horizontal pumping tests. In cases of that only one observation well at a distance, *r*, from the pumping well is available, the well discharge can be formulated as:

$$W = \frac{2\pi KD\left(s_w - s_1\right)}{\ln\left(r_1 / r_w\right)} \qquad (4)$$

where, s_w and s_1 are the respective steady-state elevations of the water levels in the piezometers; r_1 is the distance between pumping well and observation well; r_w is the radius of the pumping well [24].

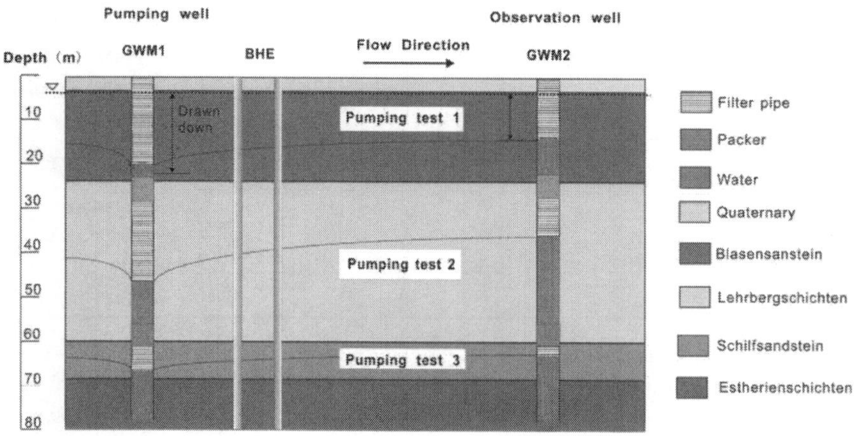

Figure 2. Schematic diagram of the pumping tests, the horizontal hydraulic properties of the geological layers have been investigated.

Hydraulic properties of the geological layers calculated using Thiem's method are listed in Table 2. The hydraulic conductivity of the

"Blasensanstein" layer consisting of sandstone shows the biggest value of 1.71^{-5} m/s. That of the middle "Lehrbergschichten" layer which interbeds sandstone and claystone is determined to be 1.27^{-5} m/s. The hydraulic conductivity of the "Schilfsandstein" layer consisting of sandstone is measured to be 9.8^{-6} m/s. Furthermore, a permeability test was conducted to measure detailed material parameters of the layers. The hydraulic conductivity of the "Estherienschichten" layer that consists of claystone is measured to be 6.74^{-9} m/s. Therefore, this layer is treated as an aquiclude.

2.1.3. System Configuration

In order to investigate the thermal efficiency, in this study, the BHE was designed with three drillhole diameters using a commercial computer program Earth Energy Design (EED) [25]. By coupling into a GSHP system of an office building in Nuremberg city of Germany, the BHE was used for space heating as well as cooling. The office building, constructed in 2008, has three floors and one basement with a total area of 1530 m^2. The GSHP system was finished the installation works and started operation in 2009.

A schematic diagram of the installed GSHP system is depicted in Figure 3. Briefly, the GSHP system consists of a BHE, water-to-refrigerant heat exchanger and indoor units. The BHE was constructed with 18 boreholes that can be classified into blocks according to the drillhole diameter: block I of 121 mm, block II of 165 mm and block III of 180 mm, as shown in Figure 4. Geometrically, the BHE was a 2 × 9 array with a distance of 6 m between adjacent drillholes. In summer, the system operates in cooling mode and the heat obtained from the building is dissipated to ground. Whereas, in winter the GSHP system extracts heat from the ground to warm up the conditioned space using heat pump (SWP 75 I, Uponor GmbH, Hassfurt, Germany). The operation of this GSHP system either in heating mode or in cooling mode was controlled by a three-way valve.

Table 2. The hydraulic properties estimated by pumping tests and permeability tests.

Geological layer	Material type	Depth (m)	Transmissivity (m^2/s)	Hydraulic conductivity (m/s)
Quaternary	sand	0–4	-	-
Blasensanstein	sandstone	4–25	3.4^{-4}	1.71^{-5}
Lehrbergschichten	Sandstone, claystone	26–55	3.8^{-4}	1.27^{-5}
Schilfsandstein	sandstone	56–62	6.88^{-5}	9.8^{-6}
Estherienschichten	claystone	62–80	-	6.74^{-9}

The connection between the BHE and indoor units are the heat pump and the plate heat exchanger. The usage of the heat pump and the plate heat exchanger depends on the system operation mode. In winter, the system operates in heating mode and the heat pump is used for space heating. On the other hand, in summer, the system runs in cooling mode. A metal plate with a high thermal conductivity is used for transfer the energy from the fluid circulating in U-tube to the building. These components and a computer server were installed in the basement, covering an area of 380 m^2. The functional area of the building (offices, meeting rooms, corridors) are located in the three above-ground floors, covering a total air-conditioned zone of 1150 m^2. Coils were embedded into the floors and heat radiators were installed in the rooms to meet different temperature requirements (*i.e.* coils were designed to keep a constant temperature of 21 °C, heat radiators is 25 °C). The operation mode of the GSHP system depends on the actual energy demand of the building.

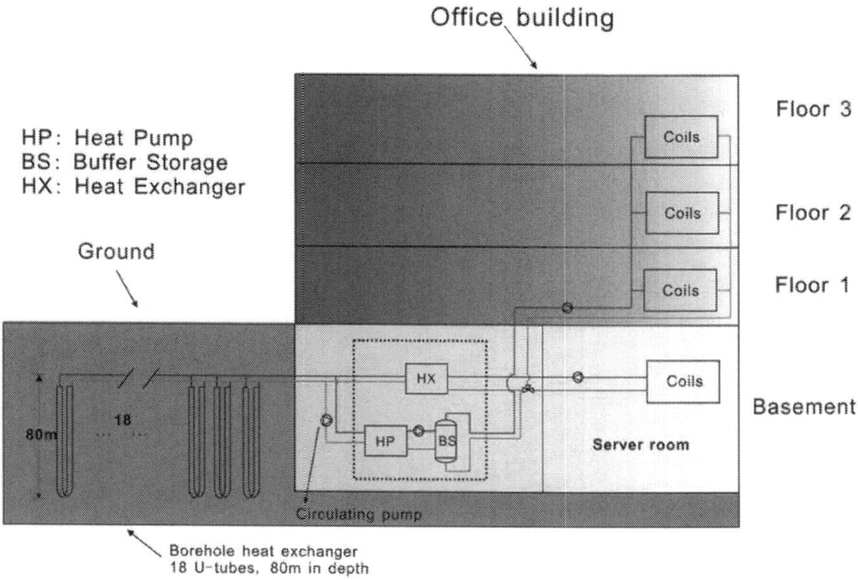

Figure 3. Schematic diagram of the GSHP system with vertical double U-tube BHEs. The BHEs of this GSHP system are in parallel connection.

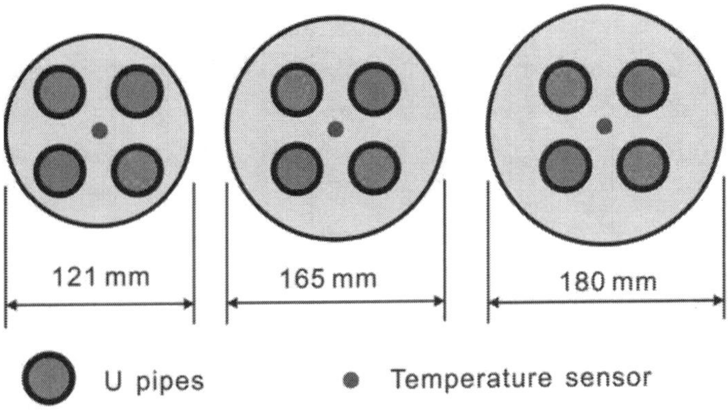

Figure 4. Schematic cross-section of the installed three-block BHEs. The temperature sensors were installed at the center of the BHEs and they are separately located at 18 m, 46.5 m and 77 m below the ground surface.

The sensors for measuring the subterranean borehole temperature are separately installed at 18 m, 46.5 m and 77 m below the ground surface. To keep the sensors at the same position, the cross-section center of the BHE, plastic holders were used. These holders with a hollow center were connected with four U-tube legs and the temperature sensors were installed in the hollow to measure the underground borehole temperature.

The specifications of the BHE components, containing configuration parameters and material properties are summarized in Table 3. It is observed that the geometric configurations of the three BHEs have the same inner configurations, but with different outer diameters. High density polyethylene double U-pipe was selected as the borehole heat exchanger. The space between the double U-pipe and the ground was grouted using enhanced cement with a thermal conductivity of 2.35 W/m/K. Glycol-based refrigerant media is circulated through the polyethylene pipe by water circulating pumps (TOP-S65/13, WILO AG, Dortmund, Germany). The properties of the grouting material are obtained from the products specifications.

Table 3. Geometric configurations and material properties for the BHE.

Components	Parameters	Specification
BHE configurations	Borehole	Borehole depth: 80 m
		Borehole distance: 6 m
		Drillhole diameter: 121 mm, 165 mm and 180 mm
	U-Pipe	Pipe type: Double-U
		Pipe shank spacing: 70 mm
		Pipe outer diameter: 32 mm
		Pipe wall thickness: 3 mm
BHE properties	Grouts	Grouting thermal conductivity: 2.35 W/m/K
		Volumetric water content: 31%
		Grouting density: 1.8 g/cm^3
	U-pipe	Pipe thermal conductivity: 0.42 W/m/K
Fluid properties	Heat carrier fluid	Fluid density: 1.11 g/cm^3
		Fluid volumetric thermal capacity: 3.8 MJ/m^3/K

2.2. Measurements Setup

2.2.1 Monitoring System

In order to investigate thermal efficiency of the BHE with different drillhole diameters, in this study, the measurements were conducted separately on each of the three BHEs. The temperature sensors (TST90, Endress + Hauser Messtechnik GmbH + Co. KG, Freiburg, Germany) were installed to measure fluid inlet and outlet temperature. At the same time, flow meters (Promag 53P40, Endress + Hauser Messtechnik GmbH + Co. KG) were installed to monitor flow rate of the fluid circulating through the U-pipes.Figure 5 shows schematic diagram of the measurement setup. It can be seen that the pipes for each six BHEs that have the same drillhole diameter were assembled into a chamber. The sensors were installed inside the chambers to monitor the fluid. Consequently, the fluid temperature and flow rate had been monitored separately for each drillhole diameter.

The specifications of these sensors, together with their models and measuring accuracy are listed in Table 4. The operating conditions of the GSHP system can then be monitored using these sensors. More specifically, the installation position of the sensors is also mentioned in this Table. The detailed installations of these sensors are depicted in Figure 4, Figure 5 and Figure 6.

Figure 6 shows a snapshot of the measurement setup on site with a view of internal settings of the plastic chamber. Two manifolds were installed for each plastic chamber, the lower one was for the fluid circulating to ground and the upper one was for the fluid circulating to the building. Pipes of the BHE were separately assembled for fluid inlet and outlet, which indicate that each six BHEs are parallelly connected. Therefore, the flow rate was evenly distributed due to the same pressure distribution of each pipe. In addition, outdoor temperature was also monitored using the sensor (TST434, Endress+Hauser Messtechnik GmbH + Co. KG). Then, thermal performance under different environmental conditions was analyzed. All the information regarding temperature, flow rate and the energy budget were continuously recorded by a data acquisition system. These parameters, together with the date and the time of the day, were also stored every 10 min.

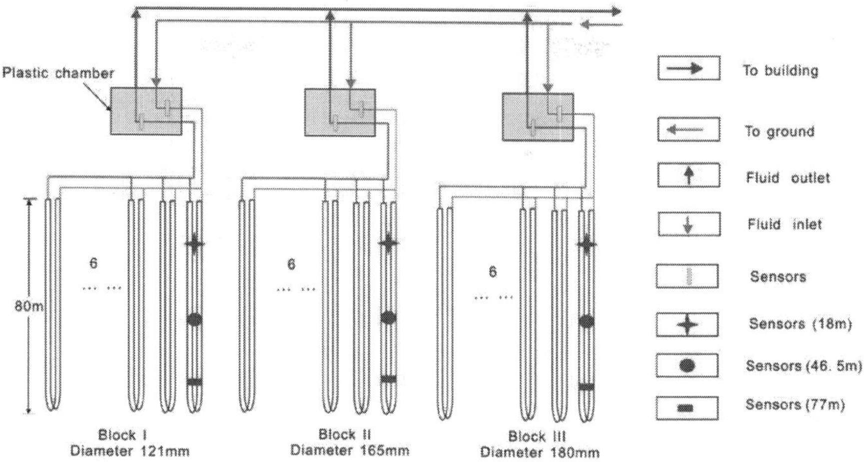

Figure 5. Schematic diagram of the measurement setup. The temperature sensors were separately installed at each block to measure the U-tube fluid in-outlet. There are twelve sensors, containing six temperature sensors and six flow meters. In addition, the temperature sensors were also installed inside the BHEs with different depths: 18 m, 46.5 m and 77 m. We select one of the sensors which have the same position in each block to measure the subterranean borehole temperature.

Table 4. The installed temperature sensors and flow meters [26].

Monitoring parameter	Components	Specification
Subsurface borehole temperature	PT-100 temperature sensors	Manufacture: Endress + Hauser Messtechnik GmbH + Co. KG Model: TST434 Measure range: −50–100 °C Uncertainty: ±0.05 °C Located at depth: 18 m, 46.5 m and 77 m
In-outlet fluid temperature	PT-100 temperature sensors	Manufacture: Endress + Hauser Messtechnik GmbH + Co. KG Model: TST90 Measure range: −50–200 °C Uncertainty: ±0.05 °C Located at fluid inlet/outlet of U-tube
Flow rate	Flow meter	Manufacture: Endress + Hauser Messtechnik GmbH + Co. KG Model: Promag 53P40, DN40 1 1/2" Assists energy: voltage 85–260 V Measure range: 0 dm^3/min–200 dm^3/min Uncertainty: ±0.2% Located at fluid inlet/outlet of U-tube
Energy budget	Data logger	Manufacture: Endress + Hauser Messtechnik GmbH + Co. KG Model: Memograph M RSG40 The results are stored every 500 microseconds

Figure 6. A snapshot of the measurements onsite. Two manifolds were installed, the lower one was for the fluid circulating to ground and the upper one was for the fluid circulating to the building. The marked 1, 2, 3, 4, 5 and 6 are for fluid circulate to the building; I, II, III, IV, V and VI are fluid circulate to the ground.

Based on the measurements, the energy exchanged between the BHE and the surrounding soil/rock at two successive entries of data logging can be calculated as:

$$E = q\rho c_m (T_o - T_i) \tag{5}$$

where E is the thermal exchange rate (kW); q is the flow rate (m^3/h); ρ is the fluid density (g/cm^3); T_i is the fluid inlet temperature (°C) and T_o is the outlet temperature (°C) [26]. The instantaneous (500 microsecond interval) heat supplied to the building was calculated by Equation (1) and the mean values with every 10min interval were stored. These values of thermal load can be positive or negative. Conventionally, the heat supplied to the building is considered to be positive.

3. RESULTS AND DISCUSSION

3.1. Typical Daily Operation

We selected August 07, which was a typical summer day, to investigate operation of the GSHP system in cooling mode. Figure 7shows the ambient temperature, fluid temperature at inlet/outlet of the ground and the corresponding flow rate in that day. The fluid inlet temperature was higher than the fluid outlet temperature, which indicates that the ground was heated by the energy removed from the building. The flow rate of the circulating fluid maintains zero when the ambient temperature dropped to 20 °C. During this period, between 01:00 AM and 09:00 AM, the GSHP system stopped operation and the fluid temperatures (inlet and outlet) were found to increase obviously. This can be attributed to the heat conduction between the circulating fluid and its surroundings at the position of BHE top where with a depth of 1.0 m. At such a depth, the soil was heated by the solar radiation during the daytime and this resulted in a high ground temperature. Therefore, the fluid inlet/outlet temperature measured at the top of the BHE was observed to increase. On the other hand, when the GSHP system was operating, the fluid temperature was almost constant, regardless of any abrupt changes in ambient temperature. The fluid inlet-outlet temperature difference was maintained at 1 °C. This fact implies that the GSHP system has a stable cooling performance in summer.

Figure 8 represents the operation conditions of the GSHP system in heating mode. The measurements were conducted on December 27, a typical winter day with the ambient temperature varying from −4 °C to 4 °C. The fluid inlet temperature is lower than the fluid outlet temperature, indicating that the ground was used as a "heat source". It is observed that the flow rate of the heat carrier fluid fluctuates frequently between zero and 4 m^3/h. At the same time, the fluid temperature fluctuates as well. Those fluctuations can be easily explained by the operation of the heat pump. In winter, the heat pump operated to extract energy from the heat carrier fluid and the fluid temperature then reduced sharply. Due to the higher ground temperature the fluid temperature can be recovered rapidly when the heat pump stopped operation. Consequently, the fluid temperature and flow rate fluctuated as observed. This finding indicates that electricity consumption of the building can also be fluctuated due to the aforementioned operation of the heat pump.

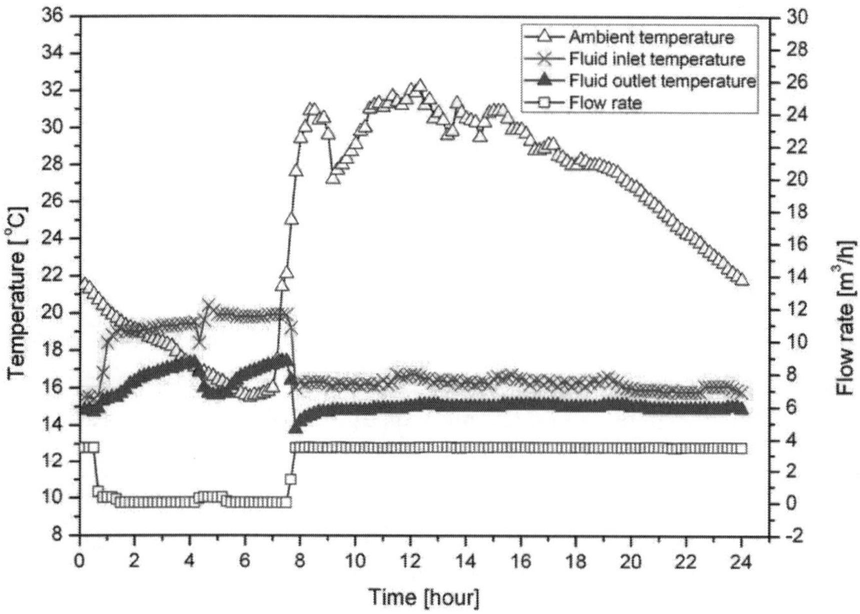

Figure 7. Recording of the GSHP system operation in a typical summer day (August 07, 2009).

3.2. Comparison of Experimental Condition

In order to compare the experimental conditions, the fluid inlet temperature of the three blocks BHEs was investigated. Figure 9represents the hourly fluid inlet temperature measured on August 07, a typical summer day. Minor difference of fluid temperature was found in the three investigated BHEs, except for the period between 01:00 AM and 08:00 AM. During this period, as shown in Figure 7, the GSHP system was shut down. The noticeable differentiation in fluid temperature was therefore caused by the heat conduction between soil and static fluid in the pipe as aforementioned in typical daily operation. To evaluate the deviations of the three blocks, the fluid inlet temperature was compared during the system operation. The fluid temperature differences of the investigated three blocks are quite small, within 0.1 °C. Considering that the measurement uncertainties of the temperature are ±0.05 °C, such a small difference can then be considered negligible.

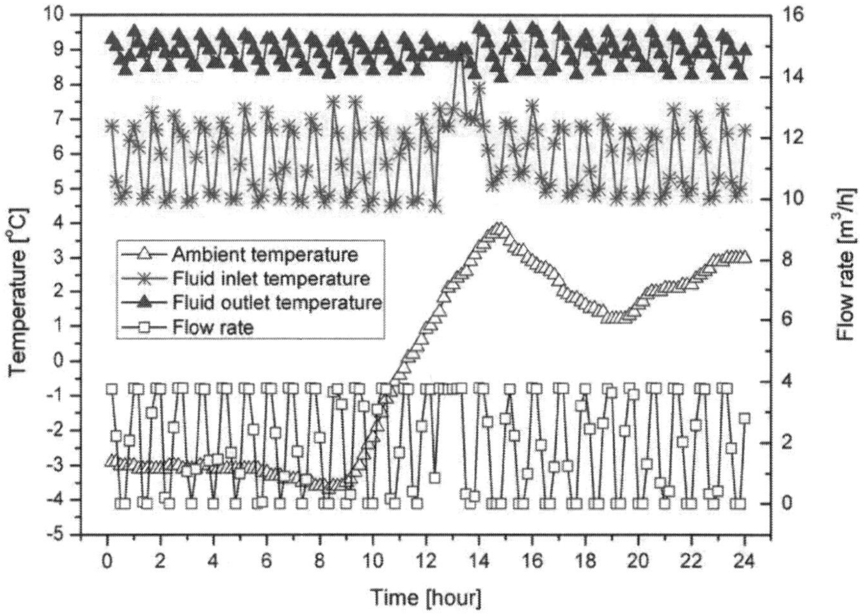

Figure 8. Recording of the GSHP system operation in a typical winter day (December 27, 2009).

On the other hand, the fluid inlet temperature was also investigated on a typical winter day. Figure 10 shows the hourly fluid inlet temperature profile on December 27, 2009. The fluid inlet temperature was observed to drop severely at the BHE of 121 mm diameter when compared to the other two blocks (block II and block III) which have a negligible difference in the fluid inlet temperature. As shown in Figure 5, the BHE of 121 mm diameter have the largest distance to the building, whereas, block II and block III have similar distances that are both smaller than that of the 121 mm diameter BHE. Considering the intermittent operation of GSHP system and the ground cooled by the chilly weather in winter, as aforementioned in Figure 8, the serious temperature drop can be well explained by heat dispersion to the ground due to the lower temperature of the surrounding soil. This finding indicates a large amount of energy loss in the horizontal pipe during winter day system operation. To achieve higher energy-efficiency, the energy loss in the horizontal pipe needs to be seriously considered in future GSHP system design and implementation in cold regions. Meanwhile, the unevenly distributed fluid inlet temperature in three BHEs ensembles suggests also that comparison of thermal performance using the data collected from winter days may deviate. Therefore, the thermal efficiency comparison, in this paper, is conducted using only the data collected during the GSHP system in cooling mode to ensure higher accuracy.

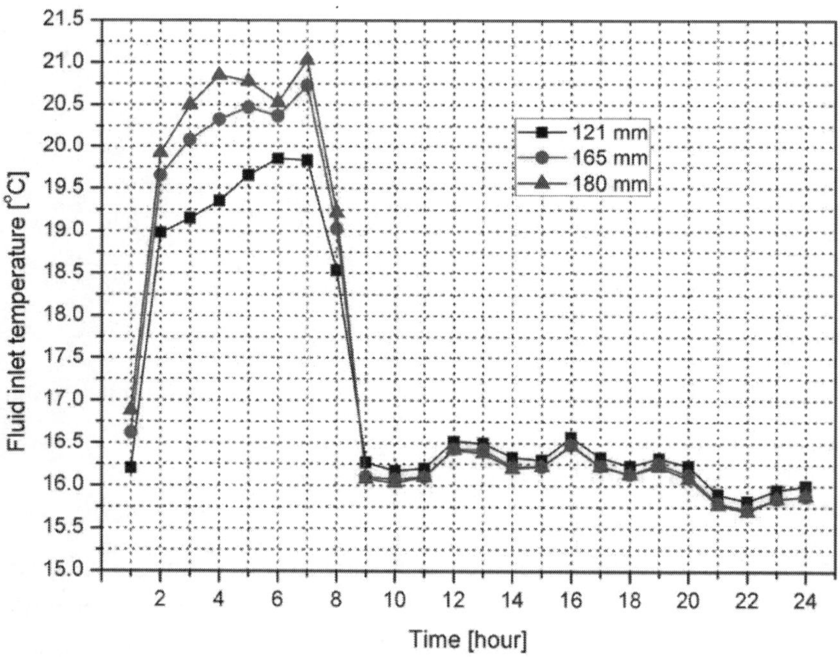

Figure 9. Hourly fluid inlet temperature distribution of the three investigated three blocks BHEs on August 07, 2009.

3.3. Comparison of Thermal Efficiency

To compare thermal efficiency more accurately, this study is solely based on data of the GSHP system operating in heating mode in 2009, the beginning of operation of the GSHP system. The data collection started from July till the end of that year. The thermal performance is estimated considering the amount of thermal exchange which can be calculated with Equation (5). Based on the thermal exchange, the thermal efficiency can then be compared as follows:

$$R_i = \frac{E_i - E_{121}}{E_{121}} \times 100\% \tag{6}$$

where R is the thermal comparison ratio (%); E is the amount of thermal exchange (kWh); and the subscript i is the object that need to be compared with the BHE with a diameter of 121 mm.

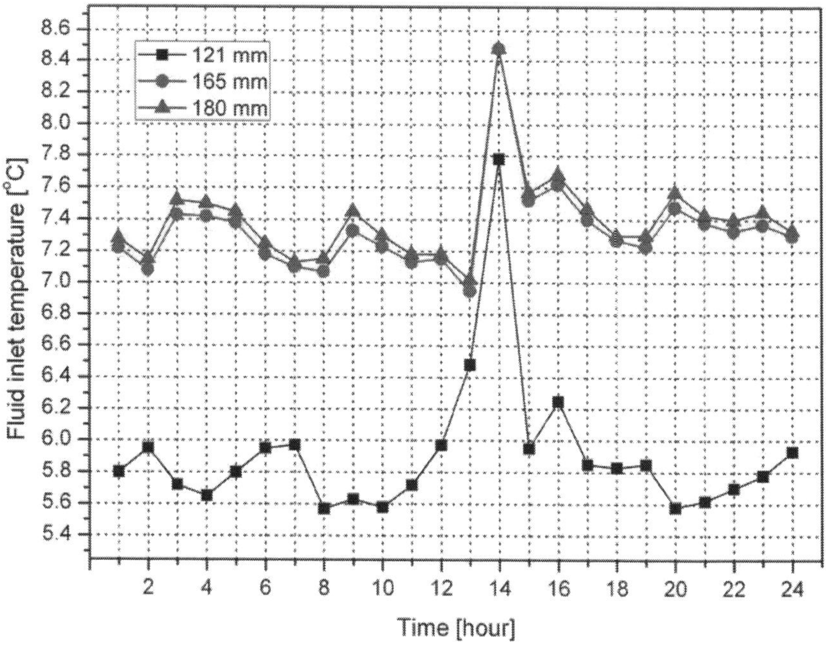

Figure 10. Hourly fluid-inlet temperature distribution of the three investigated BHEs at December 27, 2009.

Figure 11 represents the comparison results obtained using the data measured on August 07, 2009. These measurements were only conducted while the GSHP system was in operation from 10:00 AM to 05:00 PM on that day. The comparison of the thermal efficiency shows that the BHE of 180 mm diameter has the best thermal performance. On average, the daily amount of thermal exchange in the BHE of 180 mm diameter was calculated to be 6.7% larger than that the BHE of 121 mm diameter. The maximum difference between the 121 mm and 180 mm diameters BHEs was 10% in that day. For the comparison of BHE of 165 mm diameter and 121 mm diameter, the daily average ratio drops down to 2.16%, accounting for 54% of the daily maximum value (4.0%). These results indicate that increase the drillhole diameter will slightly improve the thermal performance of BHE despite the fact that the thermal conductivity of grouting material (2.35 W/m/K) is smaller than that of the ground (2.5 W/m/K).

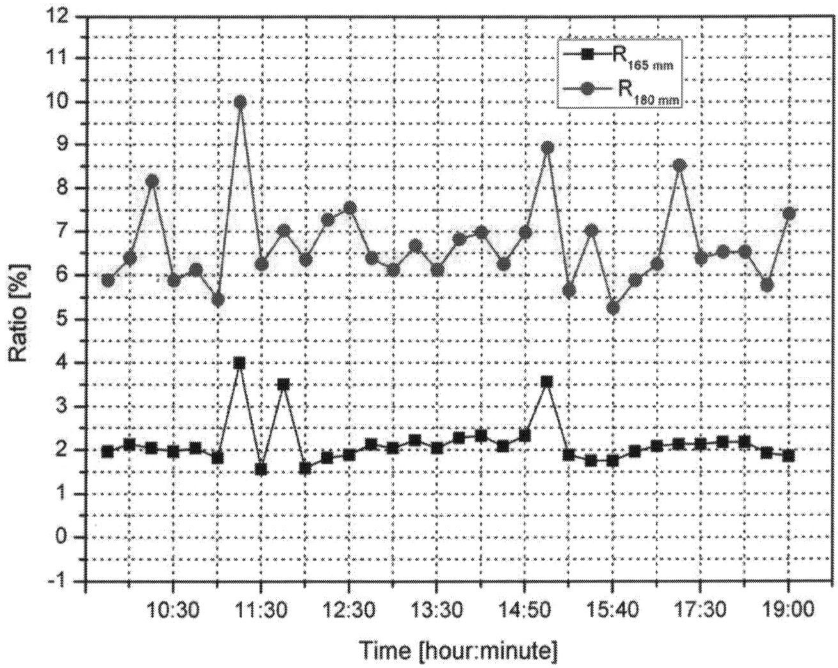

Figure 11. Thermal efficiency comparison of the BHE with three different drillholes on August 07, 2009. The thermal comparison ratio is presented for block II and block III as compared to block I.

Another fact regarding the GSHP system operation is revealed in Figure 12. As it can be seen, monthly thermal loads for the BHEs against ambient temperature are presented separately. Thermal loads of the three blocks of BHEs show an obvious increasing trend. Like the daily investigation, thermal performance of the BHE increases for larger drillhole diameter. At the same time, the thermal load of the BHEs is observed to strongly correlate with the ambient temperature. With higher ambient temperature, the cooling load increases as well.

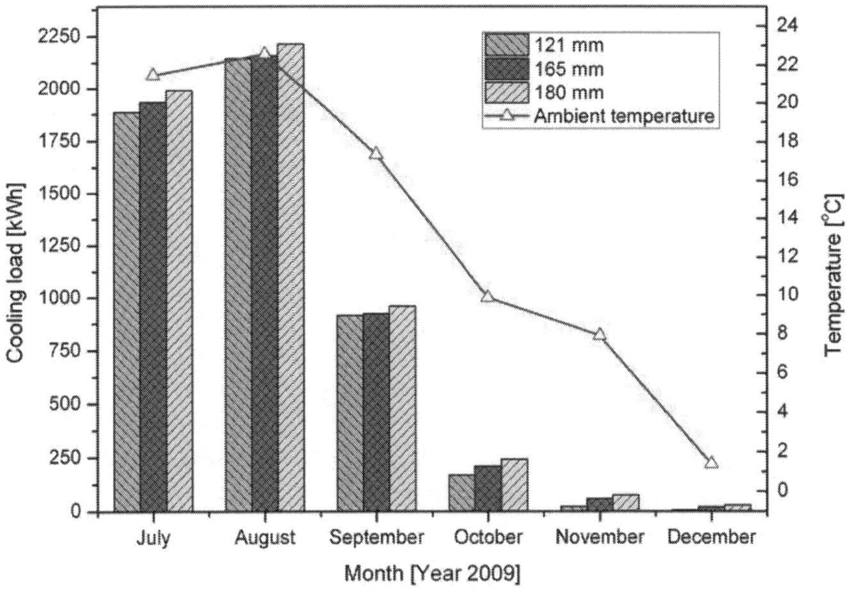

Figure 12. Monthly cooling load between July, 2009 and December, 2009.

Figure 13 shows the investigated thermal comparison ratio and the subterranean borehole temperature. The subterranean borehole temperature, in this work, was measured in the middle of the BHE, as shown in Figure 4. The mean value of the temperature measured at three depths is used due to the fact there is a negligible difference among these three depths. It is noticed that the difference of cooling performance for the BHE with different diameters is negligible when the subterranean borehole temperature reaches up to 12.5 °C.

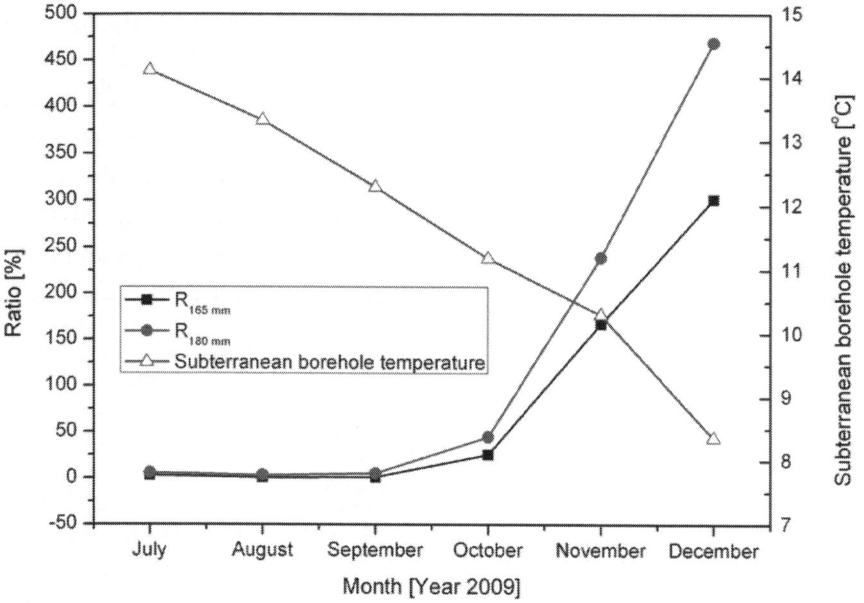

Figure 13. Thermal efficiency comparison of the BHE under different subterranean borehole temperature.

When the subterranean borehole temperature drops down to 11 °C, the difference becomes substantial. Even more, this difference will be several times larger when the subterranean borehole temperatures are lower than 10 °C. These findings suggest that thermal efficiency of the BHE with different diameters depends heavily on the subterranean borehole temperature. With a lower subterranean borehole temperature yields a higher cooling performance of the BHE with a bigger diameter.

In the long run, the thermal efficiency of the BHEs was estimated by calculating the seasonal cooling performance. Figure 14represents the seasonal amount of thermal exchange between the BHE and its surroundings. The results show an ascending trend of the amounts of heat with increased drillhole diameter. Comparing the three BHE ensembles, the amount of seasonal thermal exchange for the BHE of 165 mm and 180 mm diameter is 3.2% and 7.1% larger than that of 121 mm diameter, respectively.

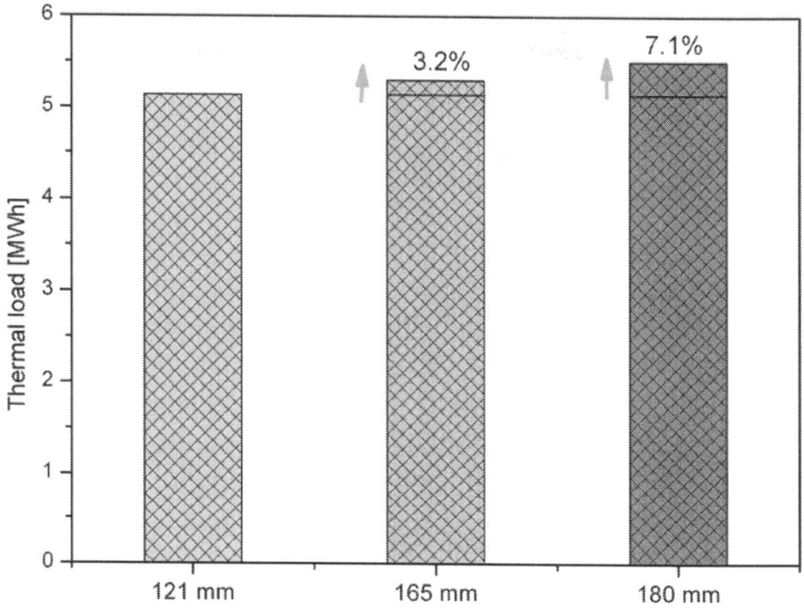

Figure 14. Comparison of seasonal thermal efficiency of BHE with different drillhole diameters.

4. CONCLUSIONS

In this paper, the thermal performance of a borehole heat exchanger (BHE) was investigated for three different drillhole diameters. This study was conducted by monitoring the operation of the GSHP system in 2009. The daily GSHP system operation was discussed separately for heating mode and cooling mode. In addition, the impacts of ambient temperature and subterranean borehole temperature on thermal performance were also analyzed. Finally, the seasonal thermal efficiency of the BHE was compared and discussed. Major conclusions are as follows:

- Thermal properties of the ground where the BHE was installed were measured by thermal response (TRT) tests. The effective thermal conductivity of the ground with a depth of 80 m was measured to be 2.5–2.6 W/m/K. This value is bigger than that of the grouting material with a thermal conductivity of 2.35 W/m/K. During performing the TRT tests, the ambient temperature was found to be an important factor to influence the testing results. Therefore, the connection pipes for a TRT test need to be carefully insulated. In addition, the hydraulic properties of the geological layers were measured by pumping tests. Based on the measurements, the geological layers were classified into different hydraulic units;

- The GSHP system was investigated for both a typical winter day and summer day. On a typical summer day, the GSHP system operates in cooling mode. On that day, the GSHP system delivers a stable cooling performance due to the constant fluid temperature and flow rate. On a typical winter day the heat pump operates to extract the energy from the heat carrier fluid, which abruptly changes the fluid temperature. Conclusively, the fluid temperature and flow rate fluctuate with time;

- In order to ensure the same experimental conditions, the fluid inlet temperature was investigated. Only negligible differences of the fluid inlet temperature were found in the three BHE blocks. However, an obvious fluid temperature drop was observed when the fluid circulates in the horizontal pipe in winter. This fluid temperature drop needs to be seriously considered in future GSHP system design and implementation in cold regions. As a result, the comparison of thermal efficiency was conducted using only the data collected during summer;

- The comparison of the daily thermal efficiency suggested that the BHE with a bigger drillhole diameter has a better thermal performance, despite the fact that the grouts have a lower thermal conductivity as compared to the ground. The investigated daily performance between 180 mm and 121 mm diameter is 6.7%, whereas this value drops to 2.16% between 165 mm and 121 mm diameter on a typical summer day. In addition, the difference of cooling performance among the three drillhole diameters depends heavily on the subterranean borehole temperature. A lower the subterranean borehole temperature yields a higher BHE cooling performance with a bigger drillhole diameter. In the long run, the amount of seasonal thermal exchange for the BHE of 165 mm and 180 mm diameter is 3.2% and 7.1% larger than that of 121 mm diameter, respectively.

ACKNOWLEDGEMENTS

This work is financed by Bayerisches Staatsministerium für Umwelt und Gesundheit. Sincere thanks are due to the Bayerische Forschung Stiftung (BFS) for providing a scholarship. The authors would also like to thank for the support provided by the Ochs Company (Nuremberg, Germany). We acknowledge support by the Deutsche Forschungsgemeinschaft and Friedrich-Alexander-Universität Erlangen-Nürnberg within the Open Access Publishing funding programme.

REFERENCES

1. Michopoulos, A.; Bozis, D.; Kikidis, P.; Papakostas, K.; Kyriakis, N.A. Three-years operation experience of a ground source heat pump system in Northern Greece. *Energy Build.* **2007**, *39*, 328–334.

2. Yang, H.; Cui, P.; Fang, Z.H. Vertical-borehole ground-coupled heat pumps: A review of models and systems. *Appl. Energy***2010**, *87*, 16–27.

3. Bose, J.; Smith, M.D.; Spiter, J.D. Advances in Ground Source Heat Pump Systems—An International Overview. In Proceedings of the 7th International Energy Agency Heat Pump Conference, Beijing, China, 19–22 May 2002; pp. 313–324.

4. Gehlin, S. Thermal Response Test: *In Situ* Measurements of Thermal Properties in Hard Rock. Master's Thesis, Lulea University of Technology, Lulea, Sweden, 1998.

5. Teza, G.; Galgaro, A.; De, C.M. Long-term performance of an irregular shaped borehole heat exchanger system: Analysis of real pattern and regular grid approximation. *Geothermics* **2012**, 45–56.

6. Hwang, Y.J.; Lee, J.K.; Jeong, Y.M.; Koo, K.M.; Lee, D.H.; Kim, I.K.; Kim, I.K.; Jin, S.W.; Kim, S.H. Cooling performance of a vertical ground-coupled heat pump system installed in a school building. *Renew. Energy* **2009**, *34*, 578–582.

7. Smith, M.D.; Perry, R.L. Borehole grouting: Field studies and thermal performance testing. *ASHRAE Trans.* **1999**, *105*, 451–457.

8. Borinaga-Treviño, R.; Pascual-Muñoz, P.; Castro-Fresno, D.; Del Coz-Díaz, J.J. Study of different grouting materials used in vertical geothermal closed loop heat exchangers. *Appl. Therm. Eng.* **2012**, *50*, 159–167.

9. Philippacopoulos, A.J.; Berndt, M.L. Influence of debonding in ground heat exchangers used with geothermal heat pumps. *Geothermics* **2001**, *30*, 527–45.

10. Pahud, D.; Matthey, B. Comparison of the thermal performance of double U-pipe borehole heat exchangers measured in situ.*Energy Build.* **2001**, *33*, 503–507.

11. Borinaga-Treviño, R.; Pascual-Muñoz, P.; Castro-Fresno, D.; Blanco-Fernández, E. Borehole thermal response and thermal resistance of four different grouting materials measured with a TRT. *Appl. Therm. Eng.* **2013**, *53*, 13–20.

12. Lee, C.; Park, M.; Min, S.; Kang, S.; Sohn, B.; Choi, H. Comparison of effective thermal conductivity in closed-loop vertical ground heat exchangers. *Appl. Therm. Eng.* **2011**, *31*, 3669–3676.

13. Desmedt, J.; Van Bael, J.; Hoes, H.; Robeyn, N. Experimental performance of borehole heat exchangers and grouting materials for ground source heat pumps. *Inter. J. Energy Res.* **2012**, *36*, 1238–1246.

14. Zeng, H.; Diao, N.R.; Fang, Z.H. Heat transfer analysis of boreholes in vertical ground heat exchangers. *Int. J. Heat Mass Transf.* **2003**, *46*, 4467–4481.

15. Shu, H.; Duanmu, L.; Hua, R. *Analysis of Selection of Single or Double U-bend Pipes in a Ground Source Heat Pump System*; Energy Systems Laboratory, Texas A&M University: College Station, TX, USA, 2006.

16. Choi, J.C.; Park, J.; Lee, S.R. Numerical evaluation of the effects of groundwater flow on borehole heat exchanger arrays.*Renew. Energy* **2013**, *52*, 230–240.

17. Wang, H.J.; Qi, C.Y.; Du, H.P.; Gu, J.H. Thermal performance of borehole heat exchanger under groundwater flow: A case study from Baoding. *Energy Build.* **2009**, *41*, 395–401.

18. Zanchini, E.; Lazzari, S.; Priarone, A. Long-term performance of large borehole heat exchanger fields with unbalanced seasonal loads and groundwater flow. *Energy* **2012**, *38*, 66–77.

19. Lazzari, S.; Priarone, A.; Zanchini, E. Long-term performance of BHE (Borehole Heat Exchanger) fields with negligible groundwater movement. *Energy* **2010**, *35*, 4966–4974.

20. Lamarche, L.; Beauchamp, B. New solutions for the short-time analysis of geothermal vertical boreholes. *Int. J. Heat Mass Transfer.* **2007**, *50*, 1408–1419.

21. Bernier, M.A. Long-term ground-temperature changes in geo-exchange systems. *ASHRAE Trans.* **2008**, *114*, 342–350.

22. Sanner, B.; Hellström, G.; Spitler, J.; Gehlin, S. Thermal Response Test—Current Status and World-Wide Application. In Proceedings of the World Geothermal Congress, Antalya, Turkey, 24–25 April 2005.

23. Pröll, M. *Durchführung von 3 Thermal Response Tests in Nürnberg*; Technical Report for Bayerisches Zentrum für Angewandte Energieforschung: Nuremberg, Germany, 2012.

24. Hellström, G. Ground Heat Storage, Thermal Analysis of Duct Storage Systems. Ph.D. Thesis, University of Lund, Lund, Sweden, 1991.

25. Blomberg, T.; Claesson, J.; Eskilson, P.; Hellström, G.; Sanner, B. EED 3.0 User's Manual. Available online: http://www.buildingphysics.com/manuals/EED3.pdf (accessed on 27 December 2012).

26. Graphic Data Manager, RSG40 Memograph M, Energy Option. Available online: http://www.de.endress.com/#product/ RSG40?open&tab=documents&filter=010.030 (accessed on 27 September 2012).

CHAPTER 8

On the sizing of a flat-panel ground heat exchanger

Marco Bortoloni[1] and Michele Bottarelli[2]

[1] Department of Engineering, University of Ferrara, Via Saragat 1, 44122 Ferrara, Italy

[2] Department of Architecture, University of Ferrara,Via Quartieri 8, 44121 Ferrara, Italy

ABSTRACT

Ground-source heat pumps are a reliable technology and may represent an efficient and cost-effective option for space heating and cooling, when the investment for ground heat exchangers is reasonable. New advanced ground exchangers have been recently proposed, showing high performances also in shallow ground; their shape has not yet been investigated in literature. In the present study, an analytical solution based on the line source method is applied for sizing a novel shape. This so-called flat-panel shape is assumed to be an equivalent slinky-coil having the same heat transfer surface per unit of trench length. As overall benchmarks, two other configurations of straight pipes disposed vertically and horizontally have been sized; all devices are supposed to work in a four lined geothermal field. The building heating requirement has been evaluated assuming a simplified lumped system and three different climate zones, defined by 2,000, 2,500 and 3,000 degree days. Then, a 2D finite-element model has been implemented to solve the transient heat conduction problem in the ground. The results of the analytical formulation and numerical simulations have been compared in terms of average temperature at the wall surface of the heat exchanger. The design minimum temperature considered by the analytical method in sizing the two straight pipe configurations and the flat-panel is accurately reproduced by the numerical model. Therefore, the slinky-coil equivalent approach followed in the analytical method for sizing the flat-panel seems to be a reliable and suitable approximation.

KEYWORDS

Horizontal ground heat exchangers, Flat-panel, Analytical method, Numerical model, Sizing comparison

1. INTRODUCTION

The European policies for energy saving in buildings and reduction of greenhouse gas emissions widely support the spread of renewable energy sources, especially in space heating and cooling. In this field, ground-source heat pumps (GSHPs) have been regarded as a reliable and profitable technology due to their high energy efficiency, when the design is compliant with local environmental conditions and building energy requirements [1, 2]. Ground-coupled heat pumps (GCHPs) are a subset of GSHPs, in which the heat transfer is performed by means of vertical and deep boreholes heat exchangers (BHEs) or horizontal and shallow ground heat exchangers (HGHEs). BHEs benefit from the relatively stable temperature in deep ground, whereas HGHEs use unsteady source/sink energy storage, related to the solar energy balance at ground surface. For HGHEs, the seasonal variation of the ground temperature may lead to unfavourable working conditions and, consequently, to an efficiency reduction. Nevertheless, ground thermal drifts are not expected after long-term operation, as reported in [3–6]. Anyway, ground heat exchangers (GHEs) are the weakest link in GCHPs, due to the low soil thermal diffusivity. To improve the heat exchange efficiency and reduce the cost for the shallow horizontal installation, new arrangements for the widespread slinky-coils installation and novel shapes of GHEs have been recently proposed [5, 6].

Because the sizing of GCHPs systems needs to consider the history of thermal loads and the ground capacity to absorb or provide heat, the ground coupling for a heat pump must be sized accurately [7]. Analytical approaches based on the line source theory and cylindrical heat transfer equations are widely used for sizing traditional GHEs installations, as reported in [8, 9] and recently provided for also by Italian regulations [10]. These methods are also useful to design some types of HGHEs adopting trench configuration, such as straight pipes and slinky-coils; an analytical procedure was proposed in [9, 10] to quickly determine the HGHE overall length for these configurations. There are still relatively few numerical solutions to the ground heat transfer problem for new GHEs shapes because of their novelty. In [11], the effect of the depth of installation and the soil thermal conductivity on the performance of three different heat exchangers (straight, helical and slinky-coil) is investigated. In [4, 12] the commercial code FEFLOW is used to simulate the energy performance of a slinky-coil, taking into account the energy balance at the ground surface and new spatial arrangements, whereas in [5] FEFLOW is implemented to make a comparison between the energy behaviour of a flat-panel and a radiator. Despite the long computational time required, several numerical studies have been carried out, because their flexibility allows implementing realistic boundary conditions, including the mass transfer to take into account the effects of the soil moisture, as shown in [13] or developing an energy balance equation at the ground surface [14]. Anyway, a method for the quick and preparatory sizing of novel shapes of GHEs is not yet available.

In the present work, the above mentioned analytical method [10] is used for sizing a novel type of HGHE, the so-called flat-panel (FP), recently developed by the University of Ferrara. The implementation has been preliminary carried out assuming the flat-panel to be a slinky-coil having the same heat transfer surface. For completeness, two standard configurations of straight pipe are considered as benchmarks. Then, all sizing has been checked by means of finite element models implemented and solved with the same boundary conditions used in the analytical procedure.

2. METHODOLOGY

This study focuses on the flat-panel (FP) sizing procedure, performed adapting the analytical method provided by [9] and adopted in [10], and checked by means of a numerical model. The analysis has been carried out only in heating mode for three different climate zones (2,000, 2,500 and 3,000 DD), and a numerical verification has been also performed for two HGHEs standard configurations with benchmark purpose. A commercial finite elements code (COMSOL Multiphysics®, V4.4) has been used for solving the heat conduction problem in a 2D symmetrical domain, adopting hourly time series as boundary conditions in order to consider the ground surface temperature and the energy requirement at the HGHE. Finally, the results have been compared in terms of temperature and energy performance, taking into account the average temperature at the HGHE wall surface.

2.1 Domains of the test cases

The analytical method reported in [10] is useful to assess the overall length of trench for common GHE shapes, arrangement and energy requirement but does not solve the problem for new shapes, such as FP. So, we have considered the FP to be equivalent to a DN20 slinky-coil having the same heat transfer surface per unit length of trench. The fictitious coil has a diameter of 1 m, equal to the FP height, and its heat transfer surface area is 0.197 m^2 for each coil. In order to match the heat transfer surface of the FP (2 m^2/m_{trench}), 10.15 coils have been considered for unit length of trench, split on two levels as a standard configuration reported in [10]; thus, every metre of FP is equivalent to 31.87 m of DN20 slinky-coil.

For completeness and benchmark purposes, two other standard HGHEs configurations have been considered, taken from those provided in [10]. Both consists of eight DN20 pipes; in the horizontally aligned case (HT), they are divided in two horizontally superposed layers (4 + 4), in the vertically aligned case (VT), they are placed vertically, as shown in Fig. 1. All the HGHEs are placed at the same average depth of 1.5 m, and the geothermal system is composed of four parallel trenches with a distance of 2.74 m between the axis of the exchangers. All pipes are supposed to be in HDPE and have a 2 mm thickness.

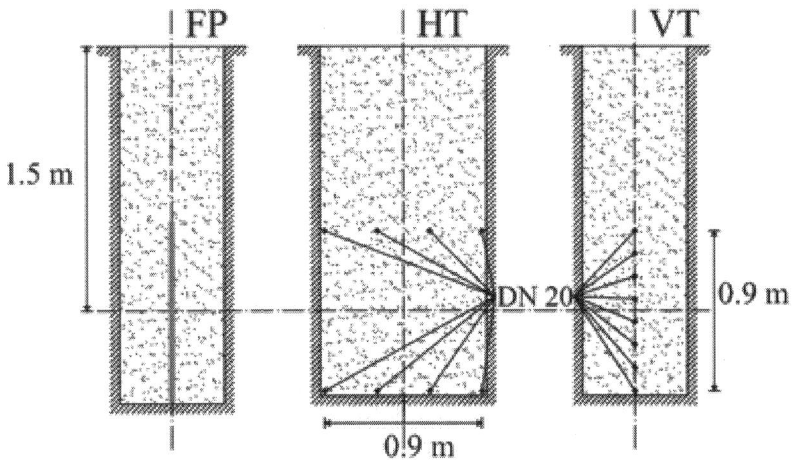

Figure 1. Cross-sectional view of HGHEs configurations (FP flat-panel, HT horizontal trench, VT vertical trench)

Our numerical analysis considers a 2D finite element domain consisting of a cross section which comprises the previous HGHEs placed within trenches and a wide surrounding soil volume. The soil is supposed to be homogenous and isotropic over the entire domain, with a thermal conductivity of 1.3 W/(m·K), a density of 1,600 kg/m^3 and a specific heat capacity of 1,200 J/(kg·K). This assumption is commonly used in literature for the purpose of modelling HGHEs. Even though heterogeneity of shallow soil may affect the results, its impact can be considered negligible, as reported in [15].

Since the overall domain is symmetric, the heat transfer problem is solved in a half-domain to reduce calculation time. Hence, only two trenches are considered for the four lined geothermal system. The computational domain is taken to be sufficiently large to have an area undisturbed by the system operation, and it is thus 14.0 m wide and 15.0 m deep, as shown in Fig. 2. Given that the temperature at the outer surface of heat exchanger has been considered, the FP is here schematized as a vertical line, while the piping as an empty circular hollow. In both cases, the thickness of the wall and its thermal resistance are neglected. Figure 2 also shows the full computational mesh and some details for each case. To reduce the computational time and the numerical errors, the grid size is fine at the ground surface and close to the GHEs, coarse for the remaining area far from it. The resulting triangular mesh is composed of up to 37,000 element, whose size ranges 17×10^{-3} cm^2 for fine grids and 10^3 cm^2 for coarse grids. The result independence from the meshing has been checked by doubling the number of the elements without relevant difference.

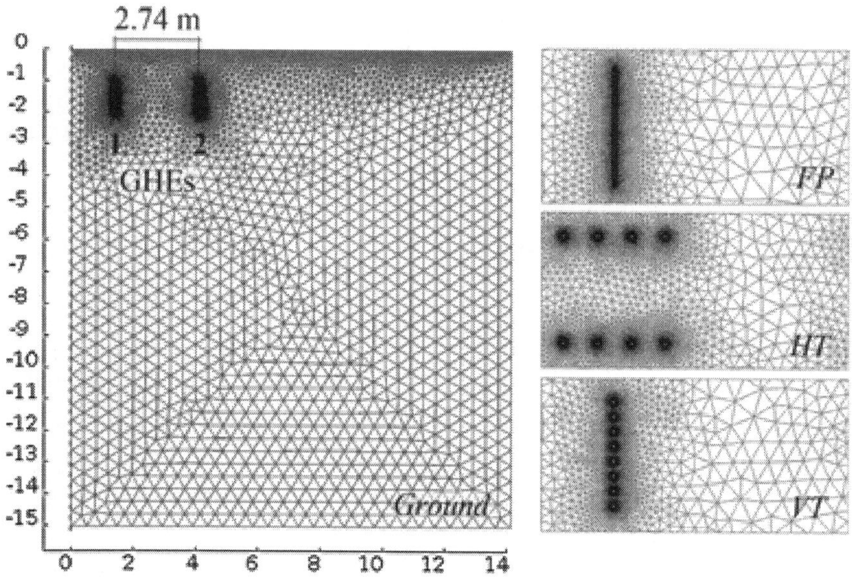

Figure 2. Computational domain, mesh details

The temperature at the GHE is then measured as the average value across its surface. Single point values for the ground temperature are calculated, between the trenches on the axis of symmetry at the average depth of the system (0.0; −1.5 m) and on an undisturbed point, 10 m far from the exchanger (14.0; −1.5 m).

2.2 Boundary conditions

According to the sinusoidal and negative exponential variation of the ground temperature reported in [10] and originally determined in [16], boundary conditions are calculated with regard to the following equation:

$$\theta_g(d,D) = \theta_M - A \cdot \cos\left[\frac{2\pi}{365}\left(D - D_0 - \frac{1}{2}\sqrt{\frac{\pi d^2}{365\alpha}}\right)\right] \cdot e^{-\sqrt{\frac{\pi d^2}{365\alpha}}}, \qquad (1)$$

where θ_g is the daily ground temperature at depth d and Julian day D, D_0 the Julian day of the lowest temperature, α the soil thermal diffusivity, θ_M the annual average air temperature and A the average annual amplitude of the air temperature.

We suppose the daily average air temperature T_{air} to be equal to the daily average surface ground temperature θ_g, as calculated by Eq. 1 setting d to zero. Then, the temperature time series on an hourly scale is obtained superimposing to the daily time series a sinusoidal oscillation ranging between the daily minimum and maximum/air temperatures (night/day) in winter and summer, taking into account the climate zone and a time drift due to the heat transfer phenomenon [17].

Specifically, in order to consider the three desired different climate zones (2,000, 2,500 and 3,000 DD) and control the heating requirement, we operate calibrating θ_M and A, using different day/night temperatures. According to data monitored at the Department of Architecture of Ferrara University, the hourly temperature at the ground surface is then smoothed by a reduction factor of 0.6 with respect to the air temperature, and a time shift of 10 h is applied between the hourly sinusoidal variations of air and ground temperature.

Table 1 reports the main parameters characterizing the climate zone supposed, while Fig. 3 shows the resulting time series for the 2,500 degree days case in winter time.

Table 1. Parameters of different climate zones

°C	2,000 DD	2,500 DD	3,000 DD
θ_M	16.49	13.99	10.75
A	12.50	12.00	11.25
T_{max} in winter	8	6	2
T_{max} in summer	34	32	28
T_{min} in winter	0	−2	−3
T_{min} in summer	24	20	16

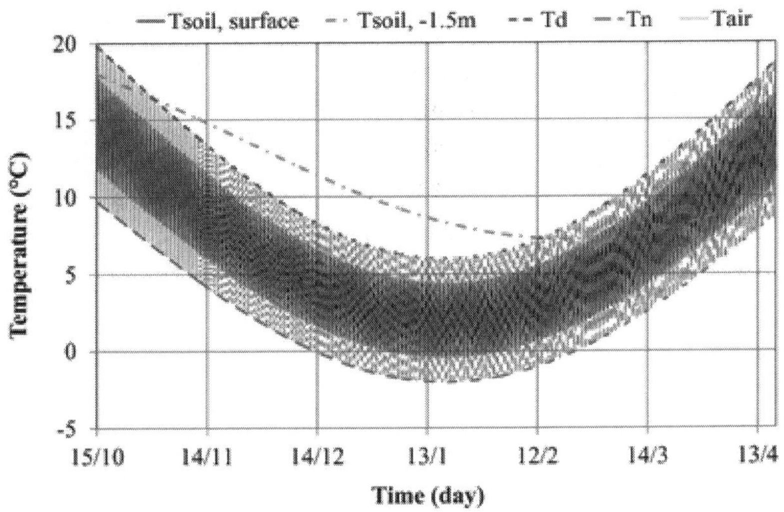

Figure 3. 2,500 DD, daily and hourly temperature time series for outdoor air and soil surface in winter time

The energy requirement for indoor space heating is defined as the amount of energy required to maintain the target indoor temperature in heating mode during the winter ($T_t = 20$ °C). For simplicity, the heating requirement is here related to the outdoor air temperature time series, assuming the building as an homogenous lumped and closed thermodynamic system, whose internal energy variation only occurs owing to the heat transfer through its envelope, as reported in [18].

As consequence, when the heating system is turned off, the average indoor temperature $T(t)$ becomes:

$$T(t) = T^{air}(t) + \left(T_0 - T^{air}(t)\right) \cdot e^{-\frac{US \cdot (t - t_0)}{r\rho Vc}}, \tag{2}$$

where T_0 is the indoor air temperature at time step t_0.

When the heating system is turned on, the specific energy requirement for unit of volume of the system can be evaluated as follows:

$$q(t) = \left[r\rho c \cdot (T_t - T_0)\right]_{t=0} + \left[U\frac{S}{V} \cdot (T_t - T_{air}(t)) \cdot \Delta t\right]_{t \geq 0}, \tag{3}$$

where the first term on the r.h.s. takes into account the energy needed to reach the target temperature (T_t) from an initial different value (T_0), and the second one the energy required to maintain it due to the heat transfer occurring through the envelope during the time step Δt, here assumed equal to one hour. Because a maximum heating power of 25 W/m^3 has been considered in the analytical method, when the resulting hourly energy exceeds the previous limit, $q(t)$ is reduced to 25 Wh/m^3 and an indoor temperature lower than the target temperature is calculated by means of Eq. 3.

According to the Italian regulations, the heating season begins on October 15th and ends on April 15th for the cases 2,500 and 3,000 DD, while it begins on November 1st for the case with 2,000 DD. The GCHPs operation hours are selected to represent actual working conditions. Cases with 2,500 and 3,000 DD involve 14 h of operation per day, from 5 am to 10 am and 4 pm to 11 pm from Monday to Friday, 8 am to 1 pm and 4 pm to 11 pm on weekends. The case with 2,000 DD implies 12 h of operation per day, from 6 am to 10 am and 5 pm to 11 pm from Monday to Friday, 8 am to 12 am and 5 pm to 11 pm on weekends. A typical week of operation for the case with 2,500 DD is shown in Fig. 4, together with the surface ground temperature and outdoor air temperature.

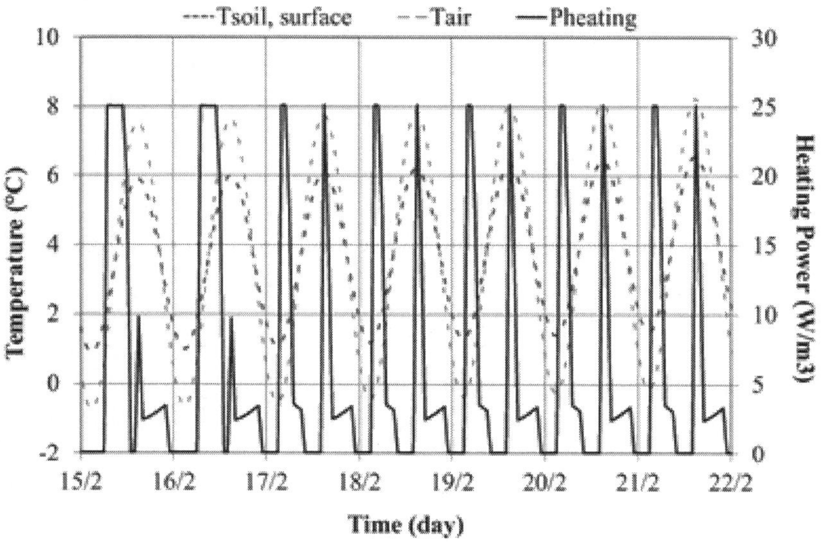

Figure 4. Weekly heating system operation

Finally, the resulting time series for the operation of the GCHPs is used to define the energy requirements at the HGHEs in the numerical model. An equivalent time series for each case is calculated to reproduce the same output temperature between the two approaches, as explained in detail in the results section.

The resulting energy requirements have been evaluated on a monthly scale for each case. The highest energy is in January for all three cases, due to the minimum value of the air temperature time series, which occurs on 15th of January according to [10]. Given an overall operation time of the GCHPs equal to 744 h in the coldest month, the specific energy requirement is 5.095, 4.968 and 5.577 kWh/m^3 for 2,000, 2,500 and 3,000 DD, respectively. The heat energy demand in January is higher for the case 2,000 DD than for the case 2,500 DD due to the lower daily operating time allowed for first one. Thus, the GCHP is forced to work for longer time at the maximum power. Consequently, it is possible to calculate the heating load factor F_h for the three cases considered. Given the maximum heating power and the total operation time, F_h is 0.274, 0.267 and 0.299 for 2,000, 2,500 and 3,000 DD, respectively.

Table 2 reports the equivalent thermal properties adopted for the simulated building.

Table 2. Thermal properties of the building

Equivalent overall thermal transmittance	U	0.3	W m^{-2} K
Shape ratio	S/V	0.65	–
Equivalent wall density	ρ	900	kg m^{-3}
Equivalent wall specific heat	c	1,200	J kg^{-1} K^{-1}
Ratio of plenum to overall volume	r	0.12	–

2.3 Analytical method

As reported in [10], the following equation defines the overall length of an horizontal heat exchanger:

$$L_{h,p} = \frac{\dot{Q}_{g,hD} \cdot \left(R_p + R_g \cdot P_m \cdot S_m \cdot F_h\right)}{\theta_{g,l} - \theta_{wD}}, \tag{4}$$

where Q˙g,hD is the required maximum heating power, $\theta_{g,l}$ is the minimum temperature of the soil at the HGHE average depth, θ_{wD} is the lowest design average temperature between inlet and outlet of the working fluid, R_p and R_g are the thermal resistances of the pipe and the soil, P_m is a coefficient related to the diameter of the pipes, S_m is the correction factor related to the distance between the trenches, and F_h is the heating load factor of the month with the highest heat requirement.

The values of R_g, P_m and S_m depend on the configuration of the heat exchanger, and are provided by tables reported in [10]. For the FP case, these are obtained through the equivalence to a slinky-coil described above, and then by interpolation. According to [10], P_m is taken equal to

1, since the diameter of the pipes is DN20, whereas R_p is here neglected, because it is two times lower than R_g. According to this principle, the average temperature of the working fluid is here taken equal to the temperature at the contact surface between soil and exchanger.

In the following Table 3, the values considered for R_g and S_m are summarized with reference to the chosen configuration, whereas the remaining parameters in Eq. 4 are detailed further.

Table 3. Parameters for the considered HGHE configurations

	Flat-panel FP	Horizontal straight pipes HT	Vertical straight pipes VT
R_g $(K\ m^2/W)$	8.44	1.80	2.48
S_m	1.21	1.21	1.15

As said, F_h represents the ratio between the HP equivalent hours of operation in heating at the maximum power and the total hours of the month concerned, as follows:

$$F_h = \frac{Q_{hD}}{\tau_{hD} \cdot \dot{Q}_{g,hD}}, \tag{5}$$

where Q˙hD is the overall energy demand in the month considered, τ_{hD} is the number of hours in the month and Q˙g,hD the maximum heating power. Equation 4 and 5 may also be considered valid for an unitary building volume, so that energy and maximum power are expressed for a single cubic metre; here, the maximum power in heating has been fixed to 25 W/m^3, a typical value for common buildings.

With regard to the hourly time series of the building heating requirement described in the following section, January proves to be the design month, due to the highest overall thermal load for the 3 climatic zones characterized, as reported in Table 4. The overall energy demand is affected by the shorter operating time for 2,000 DD (12 against 14 h/day) and it is higher than that for 2,500 DD, despite the warmer outdoor air temperature. Then more energy is required to restore the indoor comfort condition due to the longer off-period. Consequently, the resulting overall lengths calculated by Eq. 4 and reported in Table 6 are greater for 2,000 DD than 3,000 DD.

Table 4. Heating load factor in January

DD	$\dot{Q}_{hD}$ (Wh/m^3)	τ_{hD} (h)	$\dot{Q}_{g,hD}$ (W/m^3)	F_h (−)
2,000	5095.0	744	25	0.274
2,500	4968.3	744	25	0.267
3,000	5577.8	744	25	0.299

Finally, given the thermo-physical properties of the soil, it is possible to evaluate the minimum temperature of the soil at the HGHE medium depth, $\theta_{g,l}$ which occurs on February 15th. Therefore, the average temperature of the working fluid θ_{wD} is chosen to be 6 K lower than $\theta_{g,l}$ for each climate zone, according to the values suggested by [10].

In Table 5, the values for θ_{wD} and $\theta_{g,l}$ used in the analytical method are reported.

Table 5. Design values of temperature

°C	2,000 DD	2,500 DD	3,000 DD
$\theta_{g,l}$	9.5	7.2	4.4
θ_{wD}	3.5	1.2	−1.4

In order to make the analytical method comparable to the numerical model, a procedure has been developed to calculate the energy requirement at the heat exchanger for unit length of trench. From Eq. (4), the length l_h (m_{pipe}/m^3) is calculated as the overall pipe length needed to meet the energy demand of a building unit volume for each HGHEs, according the assigned specific maximum heat power (25 W/m^3), heating requirements and other parameters. Furthermore, the length of pipe available in a metre of trench, L_{HGHE}, is known for the various HGHEs configurations. It is equal to 31.87 m_{pipe}/m_{trench} for the FP case, equivalent to 10.15 coils arranged on two levels having the same FP heat transfer surface, while it is 8 m_{pipe}/m_{trench} for HT and VT. Therefore, the building heated volume per unit length of trench v_h (m^3/m_{trench}) is calculated as the ratio between L_{HGHE} and l_h.

Since the two-dimensional cross section simulated with the numerical model is equivalent to a trench length of 1 m, v_h is used as a multiplier for the time series of energy requirement at the HGHE $q(t)$, previously calculated with Eq. (3). The heating system has been set to have a maximum power of 25 W/m^3, hence the time series thus obtained is characterized by an estimated maximum heat extraction rates for unit of trenchQ_{max} (W/m).

3. RESULTS

Table 6 summarizes the values of l_h, v_h andQ_{max} adopted for every heat exchanger and climate zone considered.

The resulting heat extraction rates for FP are 67.9, 69.7 and 62.3 W/m for 2,000, 2,500 and 3,000 DD, respectively. For HT, the heat extraction rates are 78.3, 80.4 and 71.9, whereas for VT they are 60.2, 61.8 and 55.2; thus, the HT case is 15 % higher than FP, while VT is 11 % lower. Although it is not the main purpose of this work, the difference in terms of heat extraction rate between the different HGHEs should be highlighted. This is linked mainly to the design of each HGHE. FP and VT have similar geometry but VT has a lower heat transfer surface area, while HT, whose cross section is wider, is able to involve a larger volume of soil. With regard to l_h, it should be remembered that every metre of FP is equivalent to 31.87 m of slinky-coil, so that the FP lengths are 0.37, 0.36 and 0.40 m for 2,000, 2,500 and 3,000 DD, for building unit volume.

Table 6. Resulting maximum heat extraction rate of the different HGHEs

Case	DD	l_h (m_{pipe}/ m^3)	L (m_{pipe}/ m_{trench})	v_h (m^3/ m_{trench})	$\dot{Q}_{max}$ (W/ m_{trench})
FP[a]	2,000	11.725	31.87	2.718	67.9
	2,500	11.425	31.87	2.789	69.7
	3,000	12.790	31.87	2.492	62.3
HT	2,000	2.554	8.00	3.132	78.3
	2,500	2.487	8.00	3.216	80.4
	3,000	2.780	8.00	2.877	71.9
VT	2,000	3.321	8.00	2.409	60.2
	2,500	3.237	8.00	2.471	61.8
	3,000	3.620	8.00	2.209	55.2

[a] 31.87 m_{pipe} are equivalent to 1 metre of FP

The results also show that, to achieve the same heating power, the flat-panel requires a greater length of trench than the horizontal tube exchangers, and a lower trench length than the vertical ones. This is partly explained by the larger amount of soil involved in the horizontal configuration, due to the larger cross-section of HT. Despite the higher efficiency in heat transfer rate, this results in higher digging costs to build the trench. Moreover, the soil temperature at the centre of the system for the horizontal alignment drops to lower values than for the other cases. However, neglecting the thermal resistance of the pipe wall may penalize the performance of the flat-panel. The thermal resistance is significantly lower for FP than for straight pipe exchangers, due to the lower amount of material consisting the FP for equal heat transfer surface. Thus, we suppose that the flat-panel performance could be better than those of the two other cases, finally.

The values of the daily average temperature at the wall of the HGHEs are shown in Figs. 5, 6 and 7, for the different configurations and the different climatic conditions. The temperature used in the analytical method θ_{wD} is also reported as benchmark. The temperature drops to its minimum values around the middle of February, with a time shift of about one month with respect to the minimum value of air temperature. The minimum temperature with FP is only 0.3 °C lower than that evaluated by the analytical method. This minor discrepancy is observed for all the three different boundary conditions. This demonstrates that the approach followed to size a flat-panel by means of the analytical method is correct, albeit a negligible tendency to underestimate the length of the heat exchanger is visible. In the other two configurations (VT and HT), the average minimum temperature is achieved as expected. The behaviour of VT and HT is comparable, thus their relative difference is negligible.

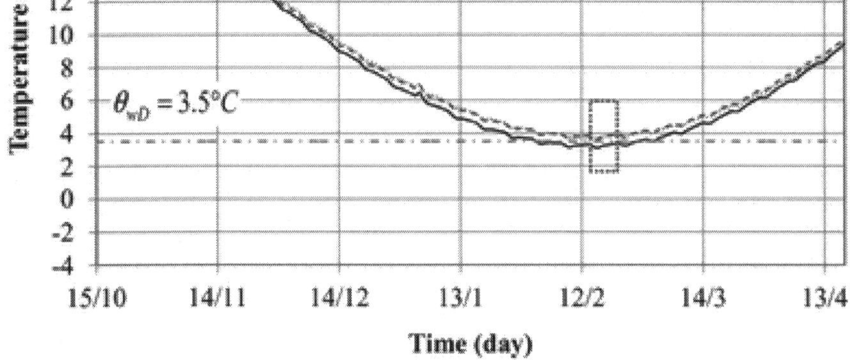

Figure 5. Daily average temperature at the HGHE surface: 2,000 DD

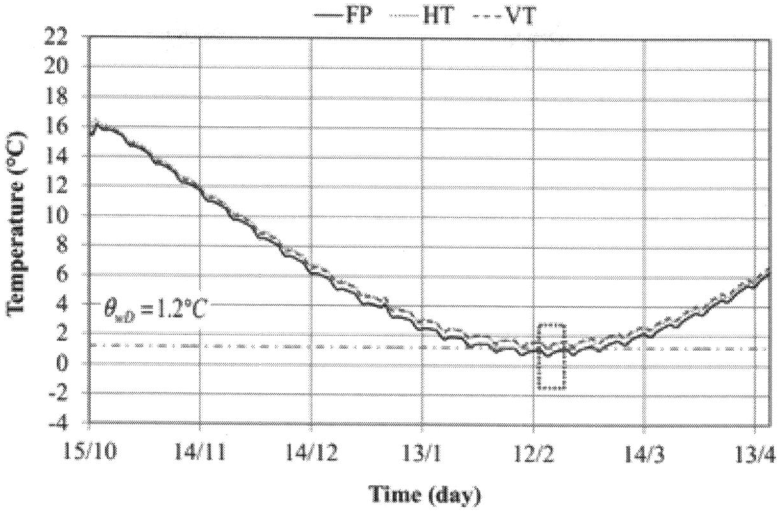

Figure 6. Daily average temperature at the HGHE surface: 2,500 DD

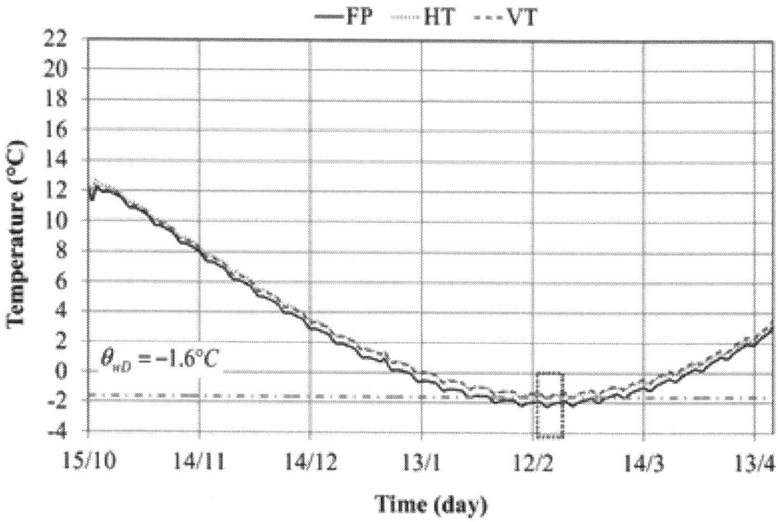

Figure 7. Daily average temperature at the HGHE surface: 3,000 DD

For completeness, a weekly detail of the hourly operation of the heat exchangers is shown in Figs. 8, 9 and10, when the minimum temperature is reached in the heating period. All the HGHEs work around at the same minimum temperature which is 2 K lower than the minimum daily average temperature. Moreover, FP displays less pronounced oscillations than HT and VT, with a lower capability of recovering. In fact, when the system is turned off the temperature in HT and VT increases of 1.3 K more than in FP.

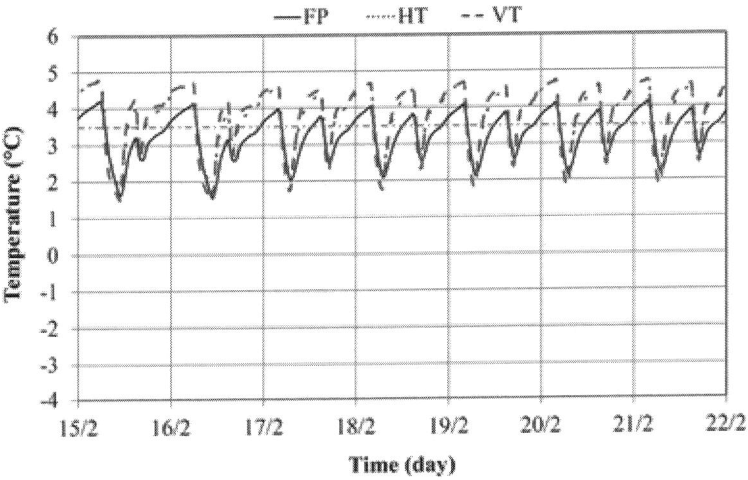

Figure 8. Hourly temperature time series at the HGHE surface: 2,000 DD

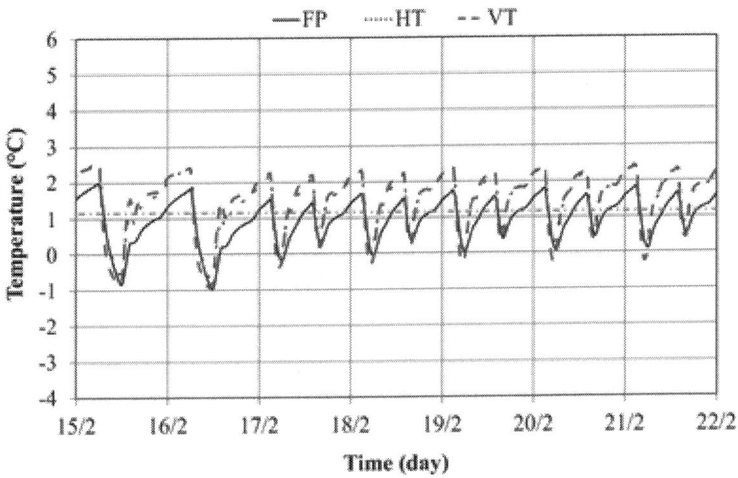

Figure 9. Hourly temperature time series at the HGHE surface: 2,500 DD

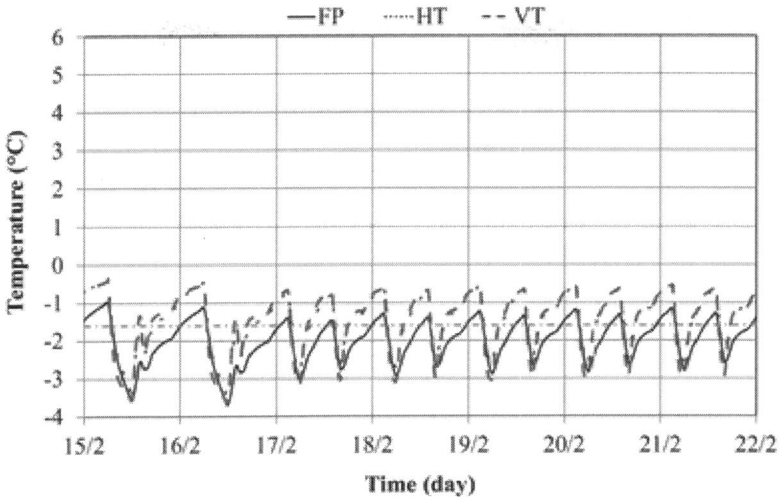

Figure 10. Hourly temperature time series at the HGHE surface: 3,000 DD

Figures 11, 12 and 13 show the ground temperature for a single measuring point placed on the domain axis of symmetry, at the average depth of the system (0; −1.5 m). The ground temperature for an undisturbed point, 10 m far from the exchanger (14; −1.5 m) is included for completeness. The amount of energy extracted by HGHE during the whole heating season is also shown. With reference to 2,500 DD, the energy exploitation made by the FP causes a decrease of the ground temperature of 3.6 °C in comparison with the undisturbed point at the same depth. The temperature drop is more pronounced, 4.1 °C, for the HT configuration, due to its higher specific power and the larger soil volume involved. On the contrary, the drop is reduced of 0.4 °C for VT. The difference between the three HGHEs remains almost unchanged for different boundary conditions (2,000 and 3,000 DD). Given the different heat extraction rate of the exchangers, the energy exploited during the entire heating season in the 2,500 DD case is 64.15 kWh/m for FP, 73.90 kWh/m and 56.84 kWh/m for HT and VT, respectively.

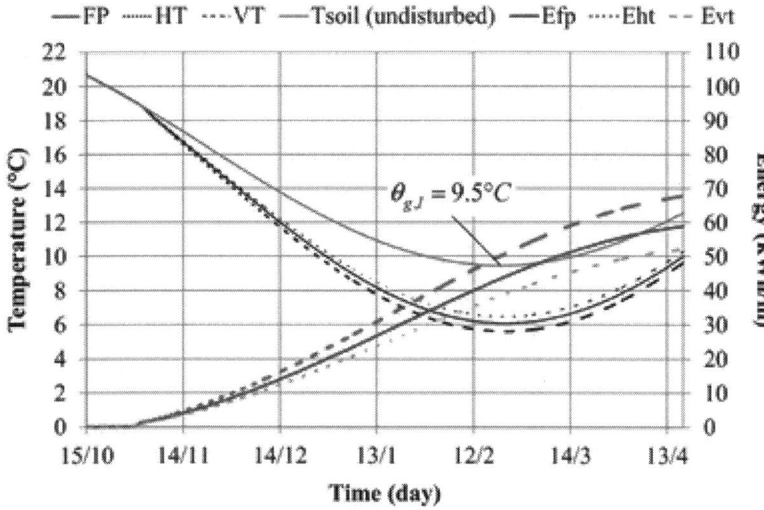

Figure 11. Temperature of the ground (0; −1.5 m) and energy extracted by HGHEs at 2,000 DD

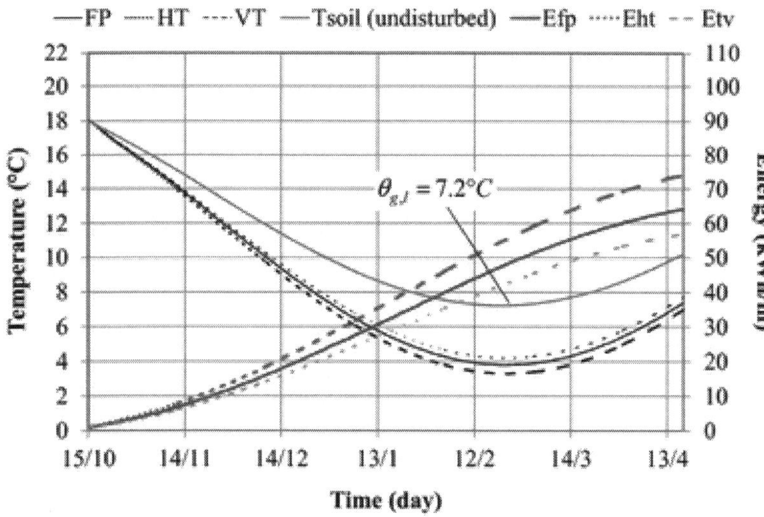

Figure 12. Temperature of the ground (0; −1.5 m) and energy extracted by HGHEs at 2,500 DD

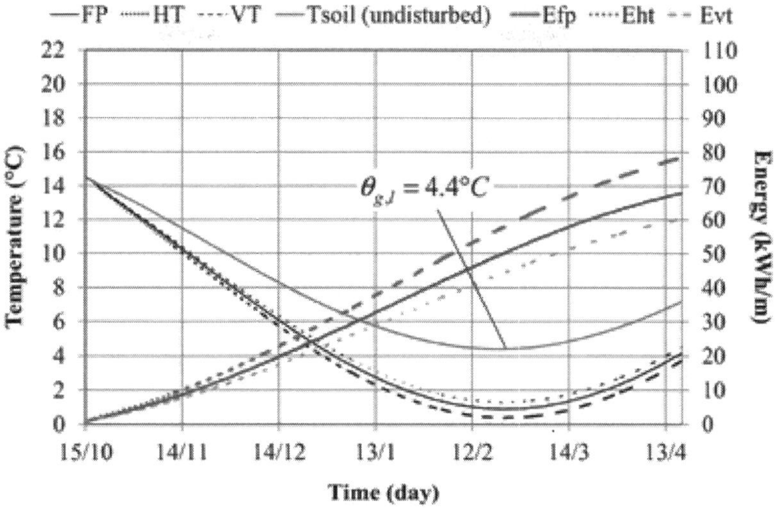

Figure 13. Temperature of the ground (0; −1.5 m) and energy extracted by HGHEs at 3,000 DD

4. FINAL REMARKS

The analytical method set by the Italian regulation [10] has been applied for sizing a novel horizontal ground heat exchanger, termed flat-panel (FP). In order to apply the analytical procedure, here the flat-panel has been considered as an equivalent slinky-coil with the same heat transfer surface per unit length of trench. Moreover, two types of vertically and horizontally aligned straight HGHEs have been considered among those established by [10] as benchmarks. Then, a 2D finite-element model has been implemented to solve the soil heat transfer problem around the ground heat exchangers thus sized, in order to check the design lengths resulting from the analytical method. The real geometry of the flat-panel has been considered in the numerical model. For simplicity, the thermal resistance of all pipe walls has been neglected.

The boundary conditions used are chosen according to the sinusoidal time series of the daily ground temperature reported in [10], which has also been extended and synchronized to the air temperature. Further detail in the forcing term has been added upon superimposing an additional sinusoidal function describing the hourly temperature variation. These boundary conditions have been applied to evaluate the heating requirements of a simplified lumped system representing a typical building.

The results of the analytical method match those of the numerical model, in terms of minimum temperature at the interface between heat exchanger and soil. It should be pointed out that the minimum temperature in FP is slightly lower than that expected; however, the resulting undersizing is negligible. Thus, the proposed approach to size a flat-panel by means of an analytical procedure has proven to be effective for each boundary condition considered.

REFERENCES

1. Mustaf, O.A.: Ground-source heat pumps systems and applications. Renew. Sustain. Energy Rev. **12**(2), 344–371 (2008)

2. Chiasson, A.D.: Advances in modeling of ground-source heat pump systems. M.Sc. Thesis, Oklahoma State University, Oklahoma (1999)

3. Gan, G.: Dynamic thermal modeling of horizontal ground-source heat pumps. Int. J. Low-Carbon Technol.**8**, 95–105 (2013)

4. Fujii, H., Yamasaki, S., Maehara, T., Ishikami, T., Chou, N.: Numerical simulation and sensitivity study of double-layer Slinky-coil horizontal ground heat exchangers. Geothermics **47**, 61–68 (2013)

5. Bottarelli, M., Di Federico, V.: Numerical comparison between two advanced HGHEs. Int. J. Low Carbon Technol. **7**, 75–81 (2012)

6. Bottarelli, M.: A preliminary testing of a flat panel ground heat exchanger. Int. J. Low Carbon Technol. **8**, 80–87 (2013)

7. Collins, P.A., Orio, C.D., Smiriglio, S.: Geothermal Heat Pump Manual. NYC Department of Design & Construction, New York (2002)

8. Kavanaugh, S.P., Rafferty, K.: Ground Source Heat Pumps-design of Geothermal Systems for Commercial and Institutional Buildings. ASHRAE Applications Handbook (1997)

9. IGSHPA: Ground Source Heat Pump Residential and Light Commercial Design and Installation Guide. Oklahoma State University, Stillwater (2009)

10. UNI 11466: Heat Pump Geothermal Systems: Design and Sizing Requirements (2012)

11. Congedo, P.M., Colangelo, G., Starace, G.: CFD simulations of horizontal ground heat exchangers: a comparison among different configurations. Appl. Therm. Eng. **33–34**, 24–32 (2012)

12. Fujii, H., Nishi, K., Komaniwa, Y., Chou, N.: Numerical modeling of slinky-coil horizontal ground heat exchangers. Geothermics **41**, 55–62 (2012)

13. Piechowsky, M.: Heat and mass transfer model of a ground heat exchanger: validation and sensitivity analysis. Int. J. Energy Res. **23**, 571–588 (1999)

14. Demir, H., Koyun, A., Temir, G.: Heat transfer of horizontal parallel pipe ground heat exchanger and experimental verification. Appl. Therm. Eng. **29**, 224–233 (2009)

15. Simms, R.B., Haslam, S.R., Craig, J.R.: Impact of soil heterogenity on the functioning of horizontal ground heat exchanger. Geothermics **50**, 35–43 (2013)

16. Carslaw, H.S., Jaeger, J.C.: Conduction of Heat in Solids. Oxford University Press, New York (1959)

17. Bottarelli, M., Di Federico, V.: Adoption of flat panels in soil heat exchange. Proceedings of 11th World Renewable Energy Congress, 330–335 (2010)

18. Bottarelli, M., Gabrielli, L.: Payback period for a ground source heat pump system. Int. J. Heat Technol.**29**(2), 145–150 (2011)

CHAPTER 9

Study of the fouling deposit in the heat exchangers of Algiers refinery

Rima Harche[1], Rafik Absi[2] and Abdelkader Mouheb[1]

[1] Laboratory of Transfer Phenomenon, Department of Chemical Engineering, Faculty of Mechanical and Process Engineering, University of Sciences and Technology Houari Boumediene (USTHB), BP 32 EL-ALIA, 16111 Bab Ezzouar Algiers, Algeria

[2] EBI, Ecole de Biologie Industrielle, Inst. Polytech. Saint-Louis,32 Boulevard du Port, 95094 Cergy-Pontoise Cedex, France

ABSTRACT

The calculation of the calender-and-tube heat exchanger is based on empirical models validated on the laboratory scale. However, when doing an extrapolation at industrial scale, errors become significant because of the non-linearities of the thermophysical properties and the transfer coefficients along the heat exchanger. For that, a follow-up on the fouling resistance in heat exchangers of tube-calender type in Algiers' refinery is presented in this study. These exchangers are used to preheat the crude oil before its passage to the atmospheric column. The results obtained from the two cells of exchangers, which were studied, showed that the resistance and the deposit fouling increases with time following an exponential curve with the existence of the fluctuations caused by the instability of the flow. Bad cleaning of the exchangers involved the absence of an induction time, and consequently, caused high values of the resistance and the deposit fouling during a relatively short time.

KEYWORDS

Heat exchanger, Calender-and-tube, Modeling, Heat transfer coefficient, Fouling resistance, Fouling deposit

1. INTRODUCTION

The accumulation of deposits on a heat transfer surface leads to a reduction of the energy transfer effectiveness [1]. The resolution of this problem requires the prevention of deposition conditions or regular cleaning of the installations. These measurements involve significant financial penalties. In order to reduce these important operations and maintenance costs, the industrials need a better understanding of this phenomenon.

There is a large variety of deposit problems and installations types prone to fouling. In this study, we present a description of the deposit phenomenon for its modeling. Tools to understand fouling must be developed to provide quantitative information for the validation of the fouling models. Several studies were devoted to the parameters influencing the fouling deposit by chemical reactions. As an example, some studies showed that the rate of fouling increases with a reduction in the fluid temperature [2]. They explain this by the fact that at constant velocity and temperature, the increase in the thermal transfer involves an increase in fouling. Others observe, on the other hand, an increase in the rate of fouling with the fluid temperature [3, 4]. Radhakrishman et al. [5] developed a predictive model using statistical methods which can a priori predict the rate of fouling and decrease in heat transfer efficiency of heat exchangers. Aminian and Shahhosseini [6] evaluated the artificial neural network (ANN) modeling, for the prediction of crude oil fouling behavior in preheat exchangers of crude oil distillation units. This research showed that the influence of the crude oil velocity and tube diameter of the fouling rate is higher than the tube surface temperature. Mohanty and Singru [7] used the C-factor for monitoring of fouling in a calender-and-tube heat exchanger. Ishiyama et al. [8] consider in their work the problem of controlling the disaster inlet temperature by using hot stream bypassing, within a PHT fouling mitigation strategy based on heat exchanger cleaning. Wang et al. [9], in their results of heat transfer experiments show that the shell-side heat transfer coefficient of the improved heat exchanger increased by 18.2–25.5 %, the overall coefficient of heat transfer increased by 15.6–19.7 %, and the exergy efficiency increased by 12.9–14.1 %. Pressure losses increased by 44.6–48.8 % with the solar installation. Sanaye and Hajabdollahi [10] used a first thermal model which is the eeNTU method which is applicable for optimal design of a shell and tube heat exchanger.

The fouling of heat exchanges used in refineries is an obstacle to their development.

In this study, we will consider the fouling phenomenon of the heat exchanger tubes for the preheat circuit of the Algiers refinery. The heat exchangers cells, E101 CBA and FED, which are used for the heating of the crude oil before its division are exposed to the problem of fouling at the tube side of heat exchangers. In this frame, a study of crude oil deposit on-site is realized which must allow to know its nature and the temporary evolution of the deposit thickness, and the most important point for this studies is to proved the linearization variation between the fouling resistance and the deposit thickness for each the two heat exchanger battery (E101 CBA and FED).

This study is composed by: a first part which presents the state of the art of fouling phenomenon, the second part presents the experimental study which was carried out in order to determine the evolution of the fouling resistance and the deposit thickness calculated based on a specific method of calculation presented in the third part, the fourth part includes our experimental results and discussions, and at the end we will finish by conclusions.

2. FOULING PHENOMENON

Fouling of heat transfer equipment is one of the most common operational problems confronted by the chemical processing industries. The main problem caused by fouling is related to the fact that the accumulation of fouling deposits has an important effect on the thermal and hydraulic performance of the affected equipment. The deposition of materials on heat transfer surfaces represents an additional resistance to the feat flow, which reduces the thermal effectiveness of the affected equipment [11].

Six different forms of fouling may occur in industrial applications, which include crystallization, particulate, chemical, corrosion, biological and solidification fouling [12].

This part considers again the various sources at the origin of this phenomenon. It is also shown how this fouling reduces the exchanger effectiveness. We will be interested in following thermal consequence of the deposit.

2.1 Origins of the fouling phenomenon

The term of fouling refers in general to any type of fouling deposit matter on the heat transfer surface. The nature of the deposit strongly depends on the composition of the fluid treated by the exchanger. The fluid can be itself at the origin of the deposit, by organic decomposition under conditions of favorable temperature and pressure. But fouling can also come from solid or semi-solid materials in suspension in the fluid.

However, all these mechanisms can be described in three stages:

1 The diffusive transport of material and its precursors through the boundary layer in close wall.

2 The adherence and the wrenching of material on the surface (clean or fouling).

3 The convective transport of this one far from the surface.

In an idealized way, Fig. 1 shows these three major stages of a fouling mechanism.

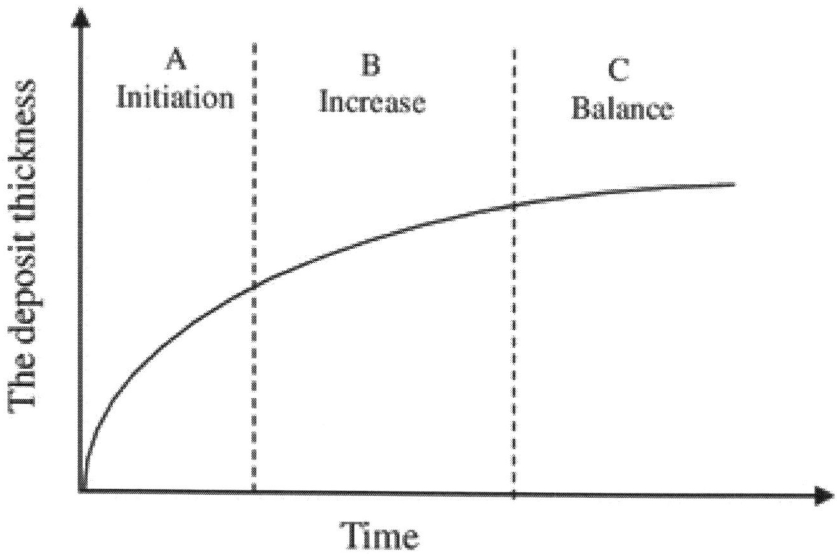

Figure 1. Three phases of a deposit mechanism: the deposit thickness according to time [4]

The first part (A) corresponds to the initiation phase. This period depends largely on the deposit type varies few minutes at a few weeks, for example in the air-conditioning systems [13]. The second part (B) corresponds to the increase in this deposit resulting from a competition between the mechanisms from deposit and wrenching. The fouling rate decreases gradually while wrenching increases, finally, leading that to a balance in (C) and a height fouling held constant.

2.2 The thermal sense of deposit

The accumulation of deposits in heat exchanger surfaces increases resistance to the heat flow. The profile of temperature under "clean" condition without deposit (index p) is shown in Fig. 2a, and under "fouling" condition (index E) in Fig. 2b. In the case of refinery, the crude oil temperature exit (T_5) is higher than temperatures T_1 (shown in figure 2) and T_5, which represent the hot gas temperatures. In the close zones wall, the fluid is slowed down and thus the heat transfer is reduced. In addition, the deposit thermal conductivity is low compared to that of metals. For these reasons, the heat transfer through the deposit requires significant differences.

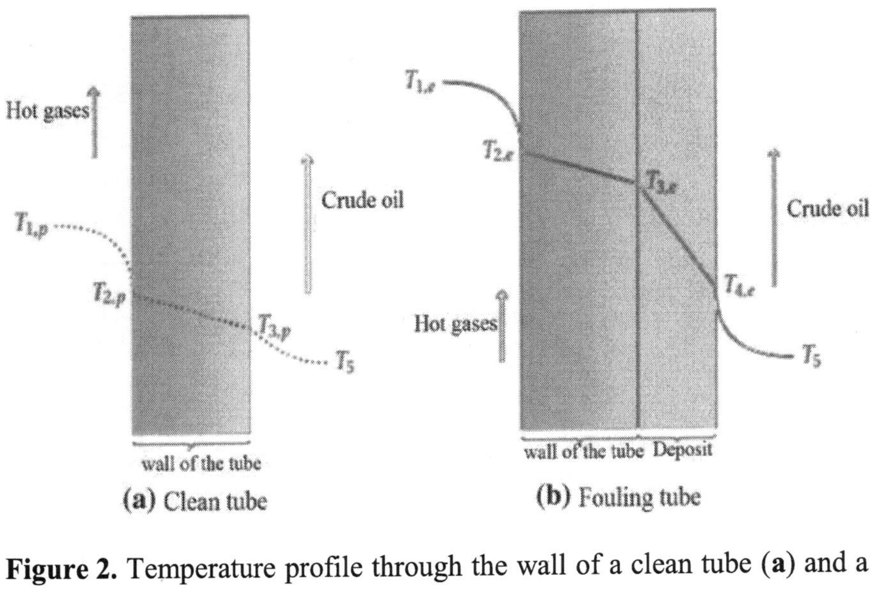

Figure 2. Temperature profile through the wall of a clean tube (**a**) and a fouling tube (**b**) in an exchanger [14]

In order to understand the fouling phenomenon, an experimental study was conducted on-site, and it is presented in the next section.

3. EXPERIMENTAL PROCEDURE

The study was made on the level of three heat exchanger cells of the circuit preheats of the Algiers refinery. The crude oil leaves the vat of storage at room temperature; one of the three centrifugal pumps P101, drives back it towards the atmospheric unit of distillation U101, while crossing the two circuits of the E101 battery E101 CBA and E101 FED, which is the tube-calenders type with a floating head. The crude crosses the battery with dimensions tube where it is heated using the backward flow head, that is a mixture of light products, coming from the top of the fractionating column C101 on the level of the plate N46. Oil passes them by the electrostatic disaster by the addition of treated water and caustic soda. Treated water is injected at the entry of the E101 exchanger and the entry of the disaster, with an aim of the crude oil washing and of involved with it salts present there (Fig. 3). Each cell is made up of three heat exchangers, laid out in a series (Fig. 4).

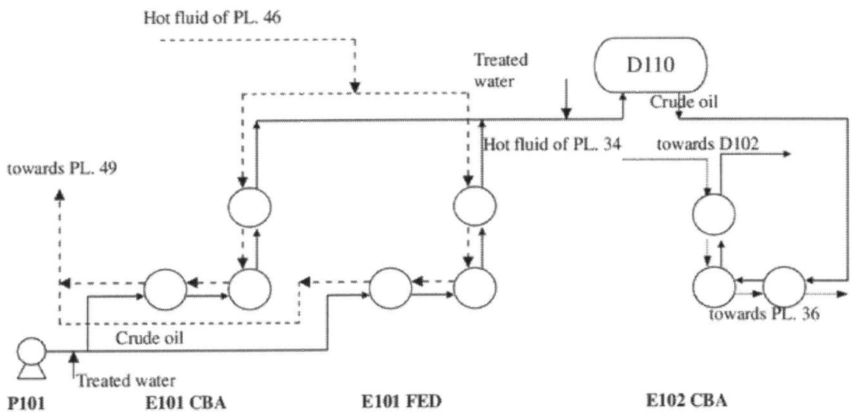

Figure 3. Simplified diagram of the crude oil preheats circuit

Our experimental study is based on the following assumptions.

1 The same crude oil flow crosses the tube side for each exchanger,

2 The same backward flow crosses the calenders side for each exchanger,

3 The number of total tubes is the sum of the each heat exchanger tube,

4 The total number of tube side master keys is equal to three times of the number of tube side master keys for each heat exchanger,

5 The output and input temperatures of the two fluids are taken at the ends of the cell.

Figure 4. Picture of the E101 CBA heat exchanger

The characteristics of these exchangers are presented in the Table below (Table 1):

We measured both cold and hot temperatures, respectively, on tube side (crude oil) and calender side (ebb of head) each 2 h from 8 to 14 h. At each time we note also the flow rate of the two fluids. At the end of the day, we obtain therefore an average temperature and flow rate values.

The calculation method, for the fouling resistance and the deposit thickness based on our results and the available data, is presented in the next section.

Table 1. Characteristics of the heat exchangers cells

Characteristics	E101 CBA FED	
	Calenders side	Tubes side
Fluid circulating	Ebb of head	Crude oil
Flow mass (lb/h)	1,000,000.00	715.260
Density API	76.00	44.40
Viscosity (cS)	–	2.40–0.96
Input/output		
Input temperature (F)	240	80
Output temperature (F)	150	220
A number of master keys	1	4
Fouling factor of	0.001	0.002
$\Delta P_{acceptable}$ (bars)	–	5
Material of construction	Carbon steel	
Calender diameter (in.)	42	
Distance between chicanes (in.)	18	
Number of calenders	6	
External diameter of tubes (in.)	0.75	
Thickness of tubes	BWG 14	
Length of tubes (in.)	20	
Total number of tubes	6,600.00	
The step: opposite square (in.)	1	
Total surface of heat exchange (ft^2)	25,000.00	
Global coefficient of heat exchange in the clean state (Btu/h ft^2 °F)	63.5	

4. CALCULATION METHOD

In the present study, the temporal evolution of the crude oil fouling resistance was studied by considering each cell consisting of three exchangers in series. Taking into account that it is not possible to measure all temperatures, we considered the three exchangers as a single one. As noticed, only one exchanger with three master keys and 12 master keys with a tube for each cell E101 CBA and FED.

The calculation of the fouling resistance was made by using the following relation:

$$R_d = \left(\frac{1}{U_s}\right) - \left(\frac{1}{U_p}\right). \tag{1}$$

U_S and U_P are the total of heat transfer surface coefficients of the dirty state and the clean state, respectively.

The total of heat transfer surface coefficients of the dirty state was given in the time course, via the expression:

$$U_s = \frac{m.c_p.(t_s - t_e)}{A.F.\,\Delta T_m}, \tag{2}$$

where m is the flow mass of the cold fluid (the crude oil), C_P is the heat-storage capacity, t_e and t_s are the output and input temperatures, respectively, A the external surface of heat transfer ($F\,\Delta T_m$) the difference in the logarithmic temperature; and U_S is the total of heat transfer coefficients brought back to external surface.

This relation is drawn from the assessment of energy on the heat exchanger by supposing the isolated system and the physical properties of the two fluids, as well as, the heat transfer coefficients remain constant along the exchanger.

Calculation of U_p

In the Algiers refinery, the operating conditions at the boundaries of the E101 exchanger are variable, it is necessary to revalue the total of heat exchange coefficients in the proper conditions U_p corresponding to the new operating conditions:

$$U_p = h_0 h_{i0}/(h_0 + h_{i0}). \tag{3}$$

With:

h_0: The heat transfer coefficient of external film.

h_{i0}: The heat transfer coefficient of internal film brought back to the external surface.

Study of the deposit thickness

The deposit thickness needs the fouling resistance and is obtained by the following correlation:

$$R_d = \frac{e_d}{\lambda_d} \quad \Rightarrow \quad e_d = R_d \,.\, \lambda_d. \tag{4}$$

However,

R_d, represents the fouling resistance;

e_d, the deposit thickness;

λ_d, the deposit thermal conductivity calculated by the Sahoo correlation (2001) [15].

For this entire calculation one uses the following fluid characteristics (Table 2):

Table 2. Fluid characteristics

Characteristics	Cold fluid Crude oil	Hot fluid Ebb of head
Density	0.7966	0.7231
Kinematic viscosity		
20.0 °C/68 °F	2.83	0.63 at 30 °C
37.8 °C/100 °F	2.02	0.57 at 40 °C
Sulphur content (% mass)	0.0634	–
Characterization factor, Kuop	12.20	–
Molecular weight (g/mol)	189	–
Asphaltene content (% mass)	0.07	–

According to our results and the data of the operating parameters obtained from the Algiers refinery (see Table 3); graphs and the deposit analyses are grouped under the title of experimental results.

Table 3. Some values of the available data.

Days	The crude oil						Ebb of head		
	t_e (°C)	t_s (°C)	Q (m^3/h)	d_4^{15}	P_e (bars)	P_s (bars)	T_e (°C)	T_s (°C)	Q (m^3/h)
E101 CBA heat exchanger									
01.04.2010	19.00	100.00	217.698	0.790	16.85	14.93	122.50	58.00	341.822
02.04.2010	19.00	100.25	204.892	0.789	16.90	15.05	122.00	57.00	326.951
05.04.2010	17.75	99.50	221.060	0.786	17.03	14.83	122.75	57.50	347.924
06.04.2010	18.00	101.50	222.415	0.791	17.00	15.08	124.50	58.25	347.269
07.04.2010	18.63	102.50	212.980	0.787	17.65	15.45	123.50	58.50	346.030
09.04.2010	19.00	102.50	217.672	0.788	17.05	15.10	123.00	59.13	349.090
12.04.2010	19.13	103.25	215.002	0.790	16.98	14.98	123.75	60.25	351.713
E101 FED heat exchanger									
01.04.2010	19.25	100.00	217.698	0.790	16.73	14.18	122.50	58.00	341.822
02.04.2010	19.00	100.00	204.892	0.789	16.93	15.50	122.00	57.00	326.951
05.04.2010	17.75	99.50	221.067	0.786	17.00	14.50	122.75	57.50	347.924
06.04.2010	18.00	100.00	222.415	0.791	16.85	14.28	124.50	58.25	347.269
07.04.2010	18,75	100.75	212.980	0.787	17.48	15.03	123.50	58.50	346.030
09.04.2010	19.00	100.00	217.672	0.788	17.68	14.40	123.00	59.13	349.090
12.04.2010	19.13	101.00	215.002	0.790	16.80	14.28	123.75	52.55	351.713

5. RESULTS AND DISCUSSION

5.1 Temporal evolution of the fouling resistance

Figure 5 shows that the temporary evolution of the fouling resistance approaches the exponential form, in spite of the deviation of some experimental points.

It is also noticed, that Fig. 5 which presents the temporary evolution of the fouling resistance for the heat exchanger battery E 101 CBA and E101 FED of Algiers refinery, seems to follow an asymptotic evolution which is in agreement with the model of Kern and Seaton [16], with the absence of the induction period. That can be explained by the delay recorded between last cleaning, and the beginning of experiments. As it appears clearly that the fouling resistance increases with time until reaching an asymptotic value which is about the 3 m^2 °C/kW for the series of exchanger E101 CBA, and about 2 m^2 °C/kW for the series of exchanger E101 FED; this is in the first part of the curve which varies at the time from 660 to 700 days, beyond the fouling resistance decreases because of the need for an emergency stop for the one exchanger of the battery E101 CBA. In the same way, it is noted that fouling resistance is higher in battery E101 CBA than in battery E101 FED; and this is right because of the position of the one compared to the other (the first battery undergoes a higher flow of crude oil and the more fouling).

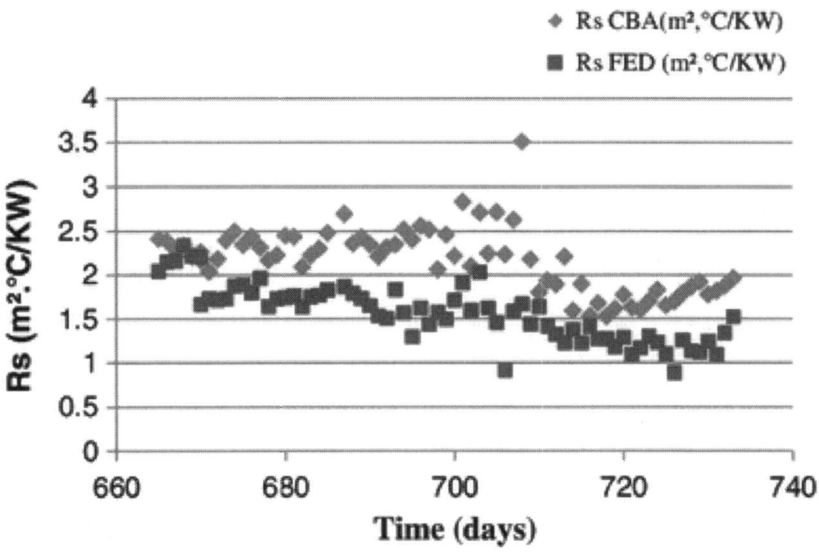

Figure. 5 Temporal evolution of the fouling resistance for the two series of exchangers

5.2 The deposit analysis

During the cleaning of the cells heat exchangers E101 CBA at the end of the study, samples of both cell deposits were taken for analysis, having used two methods.

5.2.1 Chemical analysis of the deposit

Table 4 groups the various chemical analyses made for both cell deposits.

Table 4. Chemical analysis of the heat exchangers deposits

Components	Contents E 101
Insoluble	52.44
SO_4^{2-}	18.46
Cl^-	14.20
CO_3^{2-}	00.72
Fe^{2+}	10.04
Hydrocarbons	04.07
Silica	00.07
S^{2-}	00.00
Total	100.00

These analyses let suppose that the composition of the deposit is a little unexpected. According to the Table 4, the observations which can be made are the following below:

- A content with raised insolubles.
- A low content in the hydrocarbons and in the silica.
- Of strong contents in the sulfates and in the chlorides.
- The iron is a considerable content.

5.3 Electron microscope analysis

To be able to know the deposit composition, samples of the deposit (calenders and tubes side for the two batteries of exchanger) (see Fig. 6) were analyzed under the scanning by electron microscope, which gives an exact image of the deposit structure represented in Fig. 7.

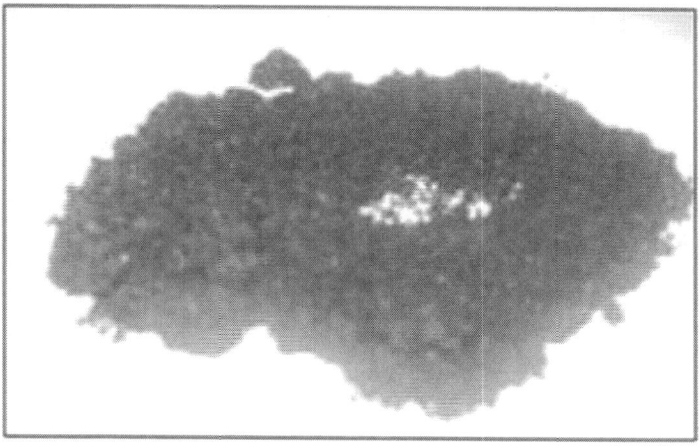

Figure 6. Samples of the crude oil deposit

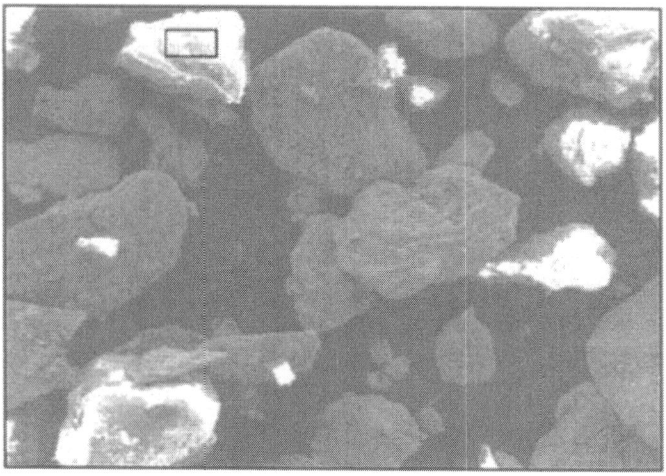

Figure 7. Picture of the sample structure of crude oil deposit analyzed by microscope

According to analyses of the crude oil deposit, the composition is given by Table 5.

Table 5. Composition of a sample of the crude oil deposit

Element	Wt%	At%
C	28.04	38.33
O	45.55	46.74
Si	24.42	14.28
S	00.20	00.10
Cl	00.00	00.00
Ca	00.21	00.09
Fe	01.59	00.47

Wt% the mass percentage, *At%* the atomic percentage

The results of the Table 5 are represented in Fig. 8 which define each element by a quite precise peak.

5.4 The deposition kinetics

All the experimental results obtained during the various taking away present the same pace as that of Fig. 9, which give the temporal evolution of the deposit thickness for two series of exchangers E101 CBA and E101 FED.

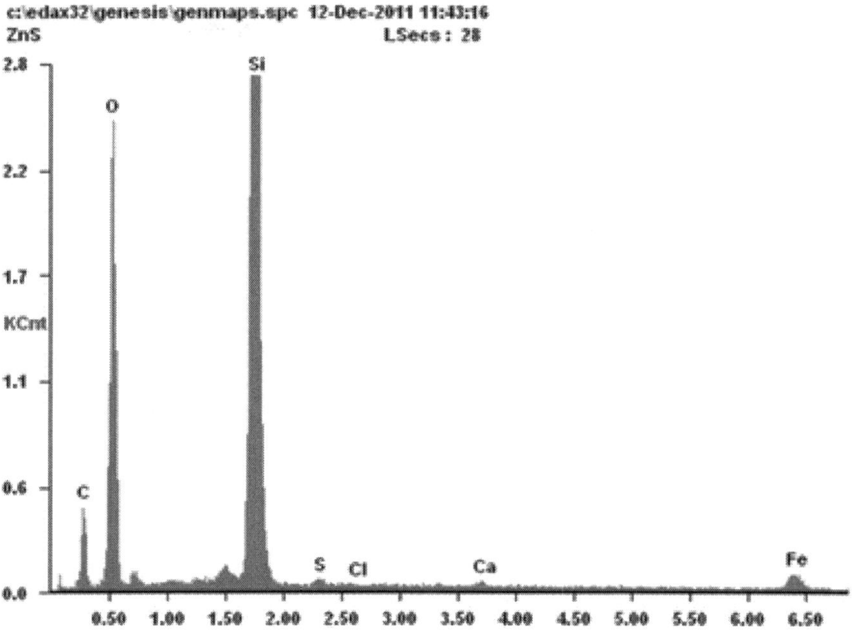

Figure 8. Energy spectrum of the crude oil deposit by the EDS technique

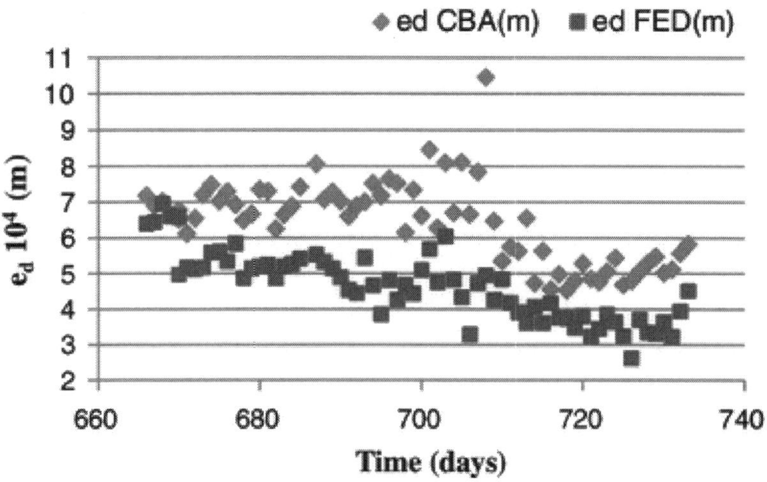

Figure 9. Temporary evolution of deposit thickness for two series of exchangers

A made observation is that the shape of the curve has an exponential evolution which agrees with the model of Kern and Seaton [16] which was already checked in the first part of the study.

Figure 9 represents two areas: firstly, the deposit thickness increases gradually until reaching a known limiting value under the name of the asymptotic value, and this in a transitory flow mode.

Additionally, in the second part of this graph, deposit thickness decreases and this because of an emergency stopping of the E101 CBA heat exchanger battery.

In the same way, it is noticed that the deposit thickness is higher in the battery E101 CBA than in the battery E101 FED; one notes that the deposit thickness varies proportionally with the fouling resistance and that is quite precise in Fig. 10, which represents the variation of the deposit thickness according to the fouling resistance for the E101 FED exchanger battery.

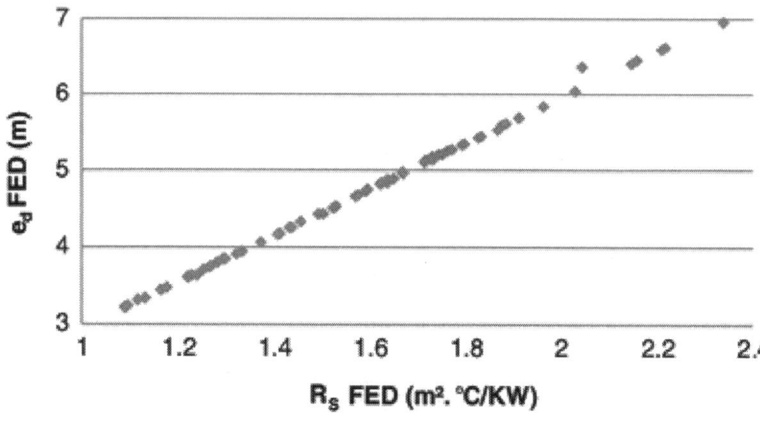

Figure 10. Variation of deposit thickness with fouling resistance for E101 FED series

6. CONCLUSION

The present study allows a better understanding, from a viewpoint of a thermo energy, of the fouling effect in the tubular heat exchangers of the preheat circuit of Algiers refinery. The study concerned the battery E101CBA and FED allowed deposit analysis, the determination of both time-dependent fouling resistance and deposit thickness and the variation of deposit thickness according to fouling resistance. The results obtained revealed an exponential evolution of the fouling resistance, in conformity with the model of Kern and Seaton with the absence of the induction period, which is explained by the consequence of a bad cleaning, or a gap between the present study and the beginning of the functioning of the heat exchangers after the last stop. We had a more important increase of the fouling resistance in the first battery E101 CBA than in the second battery E101 FED. This is obviously because the crude oil flow is higher in the first battery compared to the second battery.

During the cleaning of the heat exchangers, samples of the deposit were taken for analyses. Two methods were used:

During the cleaning of the heat exchangers, samples of the deposit were taken for analyses. Two methods were used. The first method used the chemical analysis. The results showed the coexistence of three to four types of fouling: precipitation fouling (presence of salts), particulate fouling (presence of insoluble), chemical reaction fouling (presence of hydrocarbons), and corrosion fouling (presence of iron, for example). The second method of analysis by electron microscope scanning, allows to find the same types of fouling and this confirms the former results. Finally, we proved the linear relation between the deposit thickness and the fouling resistance for each series of the exchanger battery.

REFERENCES

1. Pedot Th (2012) Modélisation du couplage thermique entre la combustion et l'encrassement des tubes d'un four de raffinerie. Ph.D. Thesis. University of Toulouse

2. Srinivasan M (2005) Heat exchanger fouling of some Canadian crude oils. Heat Transf Eng 26:7–14

3. Asomaning S (1997) Heat exchanger fouling by petroleum asphaltenes. Ph.D. Thesis. University of British Columbia

4. Saleh Z, Sheikholeslami R, Pwatkinson A (2005) Fouling characteristics of a light Australian crude oil. Heat Transf Eng 26:15–22

5. Radhakrishman VR, Ramasamy M (2005) Heat exchanger fouling model and preventive maintenance scheduling tool. Appl Therm Eng 27:2791–2802

6. Aminiam J, Shahhosseini S (2008) Evaluation of ANN modeling for prediction of crude oil fouling behavior. Appl Therm Eng 28:668–674

7. Mohanty DK, Singru PM (2001) Use of C-factor for monitoring of fouling in a shell and tube heat exchanger. Appl Therm Eng 36:2899–2904

8. Ishiyama EM, Heins AV, Paterson WR, Spinelli L, Wilson DL (2010) Scheduling cleaning in a crude oil preheats train subject to fouling: incorporating desalter control. Appl Therm Eng 30:1852–1862

9. Wang S, Wen J, Yanzhong L (2009) An experimental investigation of heat transfer enhancement for a shell-and-tube heat exchanger. Appl Therm Eng 29:2433–2438

10. Sanaye S, Hajabdollahi H (2010) Multi-objective optimization of shell and tube heat exchangers. Appl Therm Eng 30:1937–1945

11. Rodriguez C, Smith R (2007) Optimization of operating conditions for mitigating fouling in heat exchanger networks. Chem Eng Res Des 85:839–851

12. Lei Ch, Peng Z, Day Th, Yan X, Bai X (2011) Experimental observation of surface morphology effect on crystallization fouling in plate heat exchangers. Int Commun Heat Mass Transfer 38:25–30

13. Zhi-Ming XU, Zhong-Bin Z, Yang S (2007) Costs due to utility fouling in china. ECI Symposium series, Proceedings of 7th

International Conference on Heat Exchanger Fouling and Cleaning challenges and opportunities 16, p 113–118

14. Bott T (1995) The book of: Fouling of heat exchangers: chemical engineering monographs. Elsevier, p 524 ISBN: 0080531903/9780080531908

15. Manglik RM, Bergles AE (1993) Heat transfer and pressure drop correlations for twisted tap inserts in isothermal tubes: part I—laminar flows. J Heat Trans T ASME 115:881–889

16. Butterworth D (2002) Design of shell and tube heat exchangers when the fouling depends on local temperature and velocity. Appl Therm Eng 22:789–801

CHAPTER 10

Delaware Method Improvement for the Shell and Tubes Heat Exchanger Design

Miguel Toledo-Velázquez[1], Pedro Quinto-Diez[1], Juan C. Alzelmetti-Zaragoza[2], Sergio R. Galvan[3], Juan Abugaber-Francis[1], Arturo Reyes-León[1]

[1]Applied Thermal and Hydraulic Engineering Laboratory SEPI-ESIME-IPN Professional Unit "Adolfo Lopez Mateos", México DF, México

[2]Faculty of Mechanical Electrical Engineering, Universidad Veracruzana, Veracruz, México

[3]Faculty of Mechanical Engineering, Universidad Michoacana de Sán Nicolas de Hidalgo, Edifice H, University City, Santiago Tapia 403, Col. Centro, Morelia, Michoacán, México

ABSTRACT

In this paper the Delaware Method published in 1963 is analized and upgraded with using correction factors which take into account the undesirable currents of the mean flow. However, this method presents graphically these correction factors which imply an impediment to fulfill the software calculations. Thus, the equations corresponding to the correction factor equations and a Fortran 77 numerical program were established. This system is given to explore different design alternatives in order to find the optimal solution to each proposed problem. The results of this work was a simple software that can perform calculations with the introduction of parameters depending only on the geometry of the heat exchanger, i.e., geometry, temperature and fluid characteristics eliminating the human errors and increasing the calculations speed and accuracy.

KEYWORDS

Delaware Method; Flow; Correction Factor; Heat Exchanger

1. INTRODUCTION

The method is established in the heat transfer analysis and in the pressure losses of the fluid which flows through the shell side [1] [2] .

In order to complete the heat exchanger analysis, it is essential to consider the different currents generated by the shell side flow, as is shown in Figure 1. This figure has been modified by Palen and Taborek [3] [4] in regard to the original version proposed by Tinker [5] . Here five different streams are identified side of the shell. The current B is known as the mean mixed flow and flows through the mixed flow section and the window section of the heat exchanger. This is the ideal current flowing in the shell side of a heat exchanger.

Besides of flow B, there are four more flows which are present due to the free spaces and cavities between the shell and the deflectors and they provoke a modification of the current B performance.

The different leakages and recirculation currents influence the heat flux transfer in two ways:

1) Reduce the current B resulting in a drop of the heat transfer global coefficient.

2) Modify the shell-side temperature distribution.

The Delaware method considers these effects as correction factors in the heat transfer coefficient and the loss pressure calculations.

The basic equations of the heat exchangers thermal design are:

$$Q = U_{dc} AFT_{mlc} \tag{1}$$

$$U_{dc} = \frac{1}{\frac{1}{h_{io}} + \frac{1}{h_{cc}} + R_d + \frac{e_t}{k}} \tag{2}$$

In the Equation (2), h_{cc} is the fluid convection coefficient flowing in the shell and it is obtained by Equation (3). In this equation the correction factors are included, the correction factors J_i, J_c, J_l, J_b, J_r, J_r^* and J_s, which consider the different currents shown in Figure 1.

$$h_{cc} = J_i Cp_c \left(\frac{W_c}{S_m}\right)\left(\frac{k_c}{Cp_c \mu_c}\right)^{2/3}\left(\frac{\mu_c}{\mu_{wc}}\right)^{0.14} J_c J_l J_b J_r J_s \tag{3}$$

For the hydraulic design is required to determine the fluid pressure loss through the shell side and it must be included the correction factors f_i, R_b, R_l and R_s. The corresponding calculation is made using the next equation:

$$\Delta P_c = \left\{ (N_b - 1)\left[\frac{4 f_i W_c^2 N_c}{2g\rho_c S_m^2}\left(\frac{\mu_{wc}}{\mu_c}\right)^{0.14}\right] R_b R_l + 2\Delta P_{b,i}\left(1 + \frac{N_{cw}}{N_c}\right) R_b R_s + N_b \Delta P_{w,i} R_l \right\} \tag{4}$$

It is observed that the principal problem using the Delaware Method is the determination of the eleven correction factors of which nine are obtained graphically. In the next section, the nine correction factors will be presented analytically. The original version only presents analytically the $J_s$and R_s correction factors.

2. SHELL SIDE CORRECTION FACTOR EQUATIONS

According to the Delaware method, after the geometrical parameters computations, the heat transfer and loss pressure are estimated considering the correction factors. This special case will be shown in the next section by using the appropriate Equations [6] -[8].

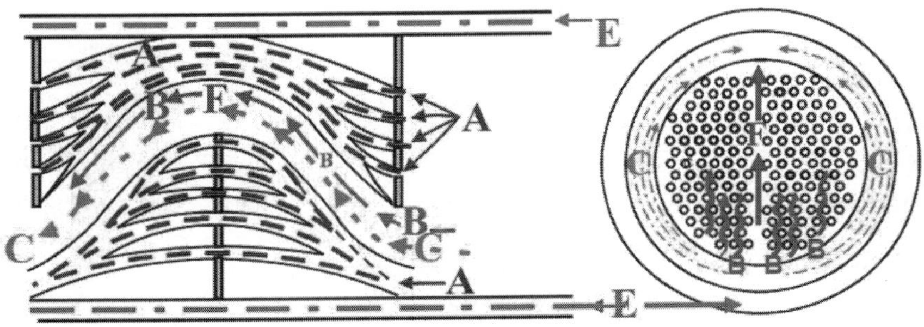

Figure 1. Ideal design of the currents flowing in the shell side. A. Currents flowing through the free spaces between tubes and baffles; B. Mean current flow; C. Recirculation current; D. Leakage current between shells and deflectors; E. Recirculation current in the past partition.

2.1. Correction Factor J_i

This factor depends on the Reynolds Number Re_c and the tube arrangement as is established in the next equation:

$$J_i = \exp\left[A + B\ln(Re_c) + C\ln(Re_c)^2 + D\ln(Re_c)^3 + E\ln(Re_c)^4 + F\ln(Re_c)^5\right]a \tag{5}$$

This equation is applied to the different arrangement tube types: triangular, square, and rhombic. For each case, the constant are presented in Table 1. It was determined 4% as that the maximum error between the equations and the graphic presentation.

2.2. Correction Factor J_c by the Baffle Configuration Effect

The deflector geometry and especially the window cross-section shape provoke currents where the effects are considered in the correction factor J_c. This factor is determined in a mathematical way as following:

$$J_c = A + BF_c + CF_c^2 + DF_c^3 + EF_c^4 \tag{6}$$

The values of each constant are presented in Table 2. Comparing the graphics obtained by Equation (6) and the original one, the maximum error established is 5%.

2.3. Correction Factor by Baffle Leakages Effect, J_l

The fluid that does not exchange heat owing to the leakage through the gap formed by the tubes and the baffles as well as through the shell and baffle is considered in the correction factor J_l. This factor is obtained using the Equation (7) where:

$$S_1 = S_{sb}/\left(S_{sb} + S_{tb}\right) \text{ and } S_2 = \left(S_{sb} + S_{tb}\right)/S_m$$

The constants required by this equation are available in Table 3. Comparing the original graph against that one getting by Equation (7), the maximum error is 2%.

$$J_l = \left[A + B(S_1) + C(S_1)^2 + D(S_1)^3\right] + \left[E + F(S_1) + G(S_1)^2 + H(S_1)^3\right](S_2) + \left[I + J(S_1) + K(S_1)^2 + L(S_1)^3\right](S_2)^2 + \left[M + N(S_1) + O(S_1)^2 + P(S_1)^3\right](S_2)^3 \tag{7}$$

Table 1. Constants for the equation for the correction factor J_i.

Arrangement	A	B	C	D	E	F
Triangular	0.627615	−0.69064	−0.0507472	0.0141049	−0.000937714	1.7683×10^{-5}
Square	0.374177	−0.671577	−0.0784051	0.02507191	−0.00224983	0.0000673254
Rhombic	−0.273166	−0.472896	−0.109701	0.023299	−0.00145983	0.0000242675

Table 2. Constants for the equation for the correction factor J_c.

For values of F_c between	A	B	C	D	E
0 and 0.9	0.533574545	0.69059596	0.290909091	−0.295959596	0
0.9 and 1	−27.84837787	152.5274893	−301.9699773	265.12743360	−86.76640715

Table 3. Constants for the equation for the correction factor J_l.

Values of S_2	A	B	C	D	E	F	G	H
0 to 0.1	1	0	0	0	−2.5903333	−4.8677761	8.773331	−7.0222218
0.1 to 0.7	0.90063003	−0.0389402	0.1363475	0.05531749	−0.4299145	−1.0949101	1.4689544	0.7341744
Values of S_2	**I**	**J**	**K**	**L**	**M**	**N**	**O**	**P**
0 to 0.1	15.8742857	50.8190476	−133.48571	106.666666	−39.6381	−168.31699	486.093325	−391.10933
0.1 to 0.7	−0.0567309	1.72685112	−4.2124254	2.4290305	0.04115536	1.3220234	3.32880672	−1.99187

2.4. Correction Factor by Recirculation Flow Effect, J_b

The recirculation currents do not exchange heat through the tubes and this issue is considered by the correction factor J_b. This factor is computed by Equation (8) which is based on F_{sbp}. The constant of the Equation $N_2 = N_{ss}/N_c$ are defined in Table 4 in terms of Re_c.

$$\begin{aligned} J_b = {} & \left[A+B(N_2)+C(N_2)^2+D(N_2)^3+E(N_2)^4\right]+\left[F+G(N_2)+H(N_2)^2+I(N_2)^3+J(N_2)^4\right]F_{sbp} \\ & +\left[K+L(N_2)+M(N_2)^2+N(N_2)^3+\tilde{N}(N_2)^4\right]F_{sbp}^2+\left[O+P(N_2)+Q(N_2)^2+R(N_2)^3+S(N_2)^4\right]F_{sbp}^3 \\ & +\left[T+U(N_2)+V(N_2)^2+W(N_2)^3+X(N_2)^4\right]F_{sbp}^4 \end{aligned} \quad (8)$$

A maximum error of 1.3% is result of comparing the original graph against that one obtained by Equation (8).

2.5. Correction Factor by Adverse Temperature Gradient J_r

This factor has the value of 1 if Re_c ³ 100. When Re_c fluctuation is between 0 and 100, the next criterion is used:

1) If Re_c £ 20, J_r will acquire the same value as J_r^* using the Equation (9), and knowing that $N_b$and $N_1 = N_c + N_{cw}$ are constants which are shown in Table 5.

$$J_r^* = \left[A + B(N_1) + C(N_1)^2 + D(N_1)^3 + E(N_1)^4\right] + \left[F + G(N_1) + H(N_1)^2 + I(N_1)^3 + J(N_1)^4\right] N_b$$
$$+ \left[K + L(N_1) + M(N_1)^2 + N(N_1)^3 + \tilde{N}(N_1)^4\right] N_b^2 + \left[O + P(N_1) + Q(N_1)^2 + R(N_1)^3 + S(N_1)^4\right] N_b^3$$
$$+ \left[T + U(N_1) + V(N_1)^2 + W(N_1)^3 + X(N_1)^4\right] N_b^4 \qquad (9)$$

A maximum error of 1.9% is result of comparing the original graph against that one obtained by Equation (8).

2) If 20 £ Re_c £ 100, J_r is computed by Equation (10) which depends on J_r^*, Re_c and Table 6constants.

$$J_r = \left(A + B\,\mathrm{Re}_c + C\,\mathrm{Re}_c^2 + D\,\mathrm{Re}_c^3\right) + \left(E + F\,\mathrm{Re}_c + G\,\mathrm{Re}_c^2 + H\,\mathrm{Re}_c^3\right) J_r^*$$
$$+ \left(I + J\,\mathrm{Re}_c + K\,\mathrm{Re}_c^2 + L\,\mathrm{Re}_c^3\right) J_r^{*2} + \left(M + N\,\mathrm{Re}_c + O\,\mathrm{Re}_c^2 + P\,\mathrm{Re}_c^3\right) J_r^{*3} \qquad (10)$$

In the previous equation, J_r^* is calculated by Equation (9). Comparing both graphics, the original and that one obtained by Equation (10), it is observed as maximum error 3.8%.

2.6. Correction Factor by Uneven Baffle Spacing at the Inlet and/or Outlet, J_s

This factor has an effect when exist a different baffle distribution at the inlet and/or outlet and along the tube bundle and it is computed by Equation (11):

$$J_s = \frac{(N_b - 1) + \left(l_{s,I}^*\right)^{1-n''} + \left(l_{s,o}^*\right)^{1-n''}}{(N_b - 1) + l_{s,I}^* + l_{s,o}^*} \qquad (11)$$

where n'' = 0.6 for turbulent flow and (Re_c > 100).

Table 4. Constants for the equation for the correction factor J_b.

For	A	B	C	D	E	F	G	H	I
$Re_c > 100$	0.99939474	0.01085613	−0.0171267	−0.1003071	0.19198207	−1.2394301	13.4729632	−85.593388	252.064457
$Re_c < 100$	0.9991889	0.02071706	−0.185339	0.59250929	−0.5964414	−1.3564474	16.400326	−116.51387	362.367998
For	J	K	L	M	N	Ñ	O	P	Q
$Re_c > 100$	−249.7121	0.74529366	−9.5329177	82.1849019	−321.67178	378.959635	−0.3575412	3.12005087	−71.95138
$Re_c < 100$	−368.18466	1.18264455	−19.614535	140.62959	−451.59784	478.689065	−1.1031648	16.9758369	−113.6298
For	R	S	T	U	V	W	X		
$Re_c > 100$	422.592705	−576.65103	0.20001675	−3.866875	70.1962129	−363.1866875	473.343386		
$Re_c < 100$	396.869404	−457.40996	0.59140536	−8.7366322	56.8613935	−210.10305	253.214998		

Table 5. Constants for the equation for the correction factor J_r^*.

A	B	C	D	E	F	G	H	I
1.11286866	−0.0444079	0.0019872	-4.324×10^{-5}	3.4787×10^{-7}	−0.0289065	2.241×10^{-4}	1.5142×10^{-6}	9.5563×10^{-8}
J	**K**	**L**	**M**	**N**	**Ñ**	**O**	**P**	**Q**
-3.11×10^{-9}	0.0008806	2.8301×10^{-5}	-2.088×10^{-6}	3.4527×10^{-8}	-9.296×10^{-11}	-1.363×10^{-5}	-9.749×10^{-7}	5.9493×10^{-8}
R	**S**	**T**	**U**	**V**	**W**	**X**		
-8.947×10^{-10}	1.618×10^{-12}	8.3072×10^{-8}	8.7699×10^{-9}	-4.727×10^{-10}	5.7282×10^{-12}	8.9289×10^{-15}		

Table 6. Constants for the equation for correction factor J_r.

A	B	C	D	E	F	G	H
−0.2477376	0.0129611	-1.127×10^{-5}	6.5427×10^{-8}	1.26556042	−0.0129666	2.5357×10^{-6}	1.5481×10^{-8}
I	J	K	L	M	N	O	P
−0.0481867	2.2364×10^{-4}	1.6928×10^{-5}	-1.708×10^{-7}	0.03112154	-2.758×10^{-4}	-6.979×10^{-6}	8.3062×10^{-8}

$n'' = 1/3$ for laminar flow and ($Re_c < 100$).

2.7. Correction Factor by Friction of an Ideal Bank Tubes f_i

The correction factor by friction in a triangular and rotated square set is determined in function of Re_c and the constants presented in Table 7 by the Equation (12).

$$f_i = \exp\{A + B\ln(Re_c) + C\ln(Re_c)^2 + D\ln(Re_c)^3 + E\ln(Re_c)^4 + F\ln(Re_c)^5 + G\ln(Re_c)^6 + H\ln(Re_c)^7 + I\ln(Re_c)^8 + J\ln(Re_c)^9\} \quad (12)$$

A maximum error of 4.9% is result of comparing the original graph to that one obtained by Equation (12).

2.8. Pressure Loss Correction Factor by Tube-Baffle Leakage R_l

The introduction of this correction factor is due to the leakage of the tube bundle and it is computed by Equation (13) which is in terms of $S_1 = S_{sb}/(S_{sb} + S_{tb})$ and $S_2 = (S_{sb} + S_{tb})/S_m$.

$$R_l = [A + B(S_1) + C(S_1)^2 + D(S_1)^3] + [E + F(S_1) + G(S_1)^2 + H(S_1)^3]S_2 + [I + J(S_1) + K(S_1)^2 + L(S_1)^3](S_2)^2 + [M + N(S_1) + O(S_1)^2 + P(S_1)^3](S_2)^3 \quad (13)$$

The constants required by the equations are established in Table 8. Comparing the original graph and that one obtained by Equation (13) is observed a maximum error of 4.5%.

Table 7. Constants for the equation for the correction factor f_i.

Type arrangement, outside diameter and pitch between tubes							
Arrangement	Diameter	Pitch	A	B	C	D	E
Triangular	19.05 mm	23.8125 mm	4.15076	−0.675323	−0.254615	0.0590471	−0.00431448
Triangular	25.4 mm	31.75 mm	4.15076	−0.675323	−0.254615	0.0590471	−0.00431448
Rhombic	19.05 mm	25.4 mm	3.69311	−1.18662	0.281179	−0.136488	0.0280708
Triangular	19.05 mm	25.4 mm	3.85004	−0.609235	−0.278939	0.0630913	−0.00452036
Rhombic	25.4 mm	31.75 mm	3.97656	−0.796179	−0.1565	0.0319655	−0.00143937
Square	19.05 mm	25.4 mm	3.76203	−0.9323	−0.0827537	0.0678788	−0.0281861
Square	25.4 mm	31.75 mm	3.99352	0.768721	−0.32634	0.176985	−0.0326325
Type arrangement, outside diameter and pitch between tubes							
Arrangement	Diameter	Pitch	F	G	H	I	J
Triangular	19.05 mm	23.8125 mm	0.000102933	0	0	0	0
Triangular	25.4 mm	31.75 mm	0.000102933	0	0	0	0
Rhombic	19.05 mm	25.4 mm	−0.00237654	0.0000713989	0	0	0
Triangular	19.05 mm	25.4 mm	0.000103903	0	0	0	0
Rhombic	25.4 mm	31.75 mm	0.00547024	−0.000459287	0.000013822	0	0
Square	19.05 mm	25.4 mm	−0.00409219	0.00245958	0.0003604	0.0000227548	-5.37988×10^{-7}

Table 8. Constants for the equation for the correction factor R_l.

Values of S_1	A	B	C	D	E	F	G	H
0 to 0.3	0.9930511	0.04163241	−0.07862612	0.01413662	−4.092978	−6.546514	4.516072	−2.525205
0.3 to 0.8	0.7499537	−0.4381533	0.2981431	0.1471556	−0.709333	0.0847061	1.053554	−4.215857
Values of S_1	**I**	**J**	**K**	**L**	**M**	**N**	**O**	**P**
0 to 0.3	15.6087	30.3289	−23.86598	14.08358	−23.22663	−41.67332	24.92778	−15.50438
0.3 to 0.8	0.3060496	1.27137	−12.85256	19.67211	−0.212271	−3.186438	18.491314	−23.16708

2.9. Pressure Loss Correction Factor by Recirculation Effect R_b

This correction factor is provoked by circulation currents in the heat exchanger and it is computed by Equation (14) which is in terms of F_{sbp} and $N_2 = N_{ss}/N_c$.

$$\begin{aligned} R_b = & \left[A + B(N_2) + C(N_2)^2 + D(N_2)^3 + E(N_2)^4\right] + \left[F + G(N_2) + H(N_2)^2 + I(N_2)^3 + J(N_2)^4\right] F_{sbp} \\ & + \left[K + L(N_2) + M(N_2)^2 + N(N_2)^3 + \tilde{N}(N_2)^4\right] F_{sbp}^2 + \left[O + P(N_2) + Q(N_2)^2 + R(N_2)^3 + S(N_2)^4\right] F_{sbp}^3 \\ & + \left[T + U(N_2) + V(N_2)^2 + W(N_2)^3 + X(N_2)^4\right] F_{sbp}^4 \end{aligned} \tag{14}$$

The constants demanded by the equation are defined in Table 9 and they are in terms of Re_cnumber. Comparing the original graph and that one obtained by Equation (14), the maximal error found is 1.3%.

2.10. Pressure Loss Correction Factor by Uneven Baffle Spacing at the Inlet and Outlet

This correction factor is proposed by the uneven baffle spacing at inlet and outlet of the heat exchanger and this factor is calculated by Equation (15).

$$R_s = \frac{1}{2}\left[\left(l_{s,I}^*\right)^{-n'} + \left(l_{s,o}^*\right)^{-n'}\right] \tag{15}$$

where n' = 1.6 for turbulent flow ($Re_c > 100$).

n' = 1 for laminar flow ($Re_c < 100$).

3. CALCULUS PROGRAM

Having the correction factors in an analytical form, it was developed a calculus program using the FORTRAN 77 language which will design quickly the shell and tubes heat exchanger through the improvement DELAWARE Method. This computing program is described next:

3.1 Flow Chart Program

The flow chart used for the shell and tubes heat exchanger is presented in Figure 2.

As from this flow chart, it was made the heat exchanger design program using the executable BELL since it runs in MS DOS platform.

4. CONCLUSIONS

This work has presented the improvements to the Delaware method for the shell and tube heat exchanger design. These improvements were made up not only of obtaining the correction factor equations which were only available in graphic forma but also of developing the computing program in Fortran 77 language.

Thus, a new and easy tool for the shell and tubes heat exchanger design is available which allows accomplishing the numerical computing in a quick form minimizing the errors of the graphical lecture. This system gives the opportunity to explore different design alternatives in order to find the optimal solution to each proposed problem.

Recent years, heat exchanger is often used for the request of technology. But the relevant design is not provided by the actual standards. This work presents the improvements to the Delaware method for the shell and tube heat exchanger design.

Table 9. Constants for the equation for the correction factor R_b.

For	A	B	C	D	E	F	G	H	I
$Re_c > 100$	0.9999956	0.02427714	−0.2498571	0.87688954	−0.9389413	−3.7854133	42.7281376	−285.42201	841.836875
$Re_c < 100$	0.9989929	0.10536115	−0.8164392	2.13589149	−1.8323906	−4.7393642	60.6001915	−402.43318	1138.26259
For	**J**	**K**	**L**	**M**	**N**	**Ñ**	**O**	**P**	**Q**
$Re_c > 100$	−823.25082	6.9400004	−114.27156	831.289197	−2493.3366	2464.67857	−7.1165478	129.6448	−995.82525
$Re_c < 100$	−1075.7794	11.8151337	−238.94173	1739.20139	−4979.2121	4724.16651	−17.614757	423.093405	−3227.2538
For	**R**	**S**	**T**	**U**	**V**	**W**	**X**		
$Re_c > 100$	3071.9304	−3083.9014	3.21390213	−60.119316	480.616728	−1521.5693	1550.23124		
$Re_c < 100$	9354.34611	−8902.6809	11.6766845	−302.46692	2376.53333	−6980.2925	6687.42629		

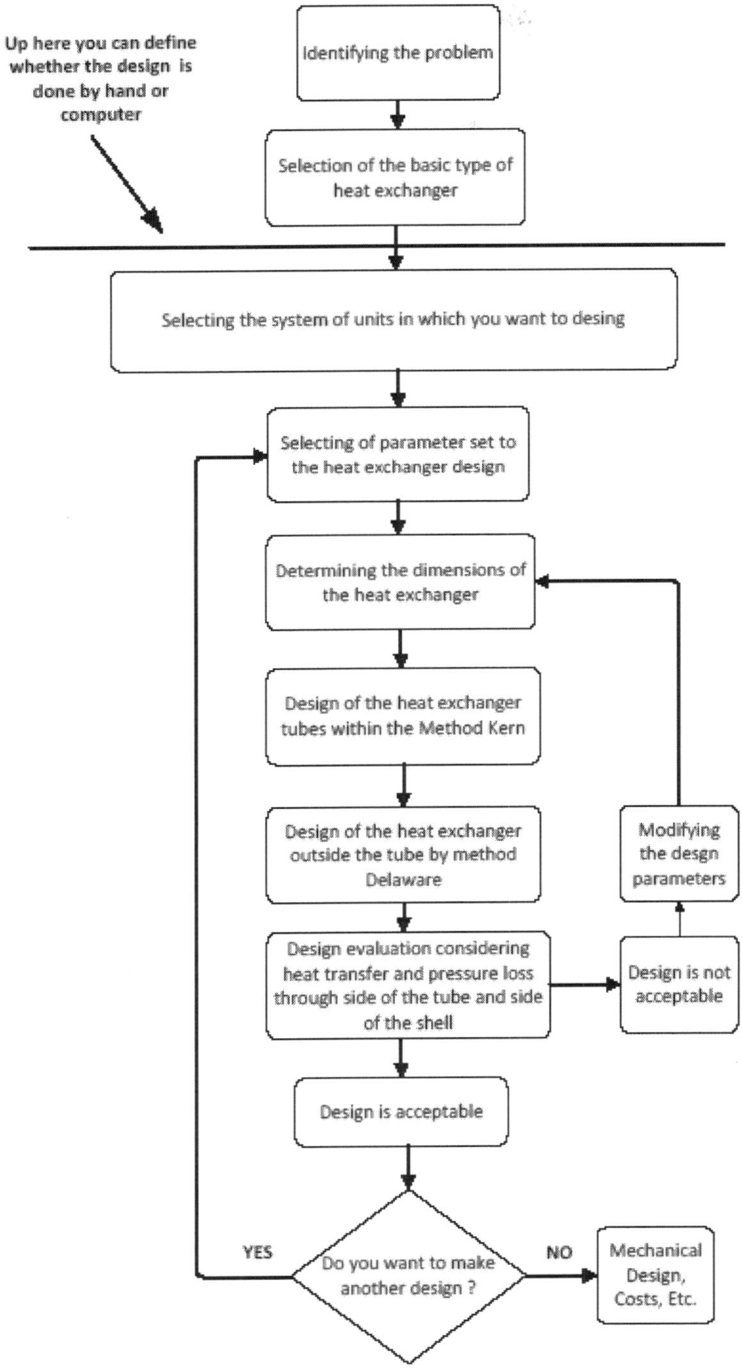

Figure 2. Flow chart program.

REFERENCES

1. Vengateson, U. (2010) Design of Multiple Shell and Tube Heat Exchangers in Series: E Shell and F Shell. Chemical Engineering Research and Design, 88, 725-736.

2. Costa, A.L.H. and Queiroz, E.M. (2008) Design Optimization of Shell and tube Heat Exchangers. Applied Thermal Engineering, 28, 1798-1805.

3. Serna, M. and Jimenez, A. (2005) A Compact Formulation of Bell-Delaware Method for Heat Exchanger Desing and Optimation. Chemical Engineering Research and Desing, 83, 539-550.

4. Shah, R.K., Dusan, P. and Sekulic, D.P. (2003) Fundamentals of Heat Exchanger Design. John Wiley & Sons, Hoboken.

5. Ayub, Z.H. (2005) A New Chart Method for Evaluating Single-Phase Shell Side Heat Transfer Coefficient in a Single Segmental Shell and Tube Heat Exchanger. Applied Thermal Engineering, 25, 2412-2420.

6. Leong, K.C., Toh, K.C. and Leong, Y.C. (1998) Shell and Tube Heat Exchanger Design Software for Educational Applications. International Journal of Engineering Education, 14, 217-224

7. León, A.R., Velazquez, M.T. and Diez, P.Q. (2011) The Design of Heat Exchangers. Engineering, 3, 911-920.

8. Castillo, R. (1999) Diseño Computacional de Intercambiadores de Calor de Coraza y Tubos por el Método Delaware. Tesis de Maestría, Instituto Politécnico Nacional, México.

CHAPTER 11

Thermal resistance of borehole heat exchangers composed of multiple loops and custom shapes

AA Koenig

College of Arts and Sciences, West Chester University, Merion Hall, West Chester, PA 19383, USA

ABSTRACT

Background

The thermal resistance of a borehole can be reduced by employing thermally enhanced grout, increasing the surface area of the loop and locating the legs proximal to the bore wall. Thermal models that are used to predict borehole heat exchange are characterized by either simplified formulations that are restrictive in their application, but utilitarian, or complex multi-dimensional analyses that are cumbersome to implement.

Methods

The borehole resistance methodology presented here offers a straightforward solution that is built on single loop conduction shape factor analysis with thermal shunt accounting and pipe-pipe configuration analysis, to extend to multi-loop borehole configurations and custom kidney extrusions.

Results

The borehole resistance predictions are compared to published data and information listed by manufacturers of multi-loop products in third party thermal tests against standard loops. The results are found to agree within the constraints posed by the model assumptions. The methodology offers a straightforward solution that can be incorporated into popular geothermal loop sizing software such as GLD, GLHEPRO and other system software.

Conclusions

The advantages and challenges of these advanced loop designs are discussed and conclusions drawn. By reducing the bore resistance, one can take advantage of less drilling and proportionally less capital cost for the bore field, while achieving the same loop temperatures.

KEYWORDS

Geothermal heating and cooling, Ground source heat pump (GSHP), Vertical loops, Borehole studies, Loop field, Multiple loops in borehole

1. BACKGROUND

One of the challenges to the ground source heat pump market is the cost burden of implementing the ground heat exchange. For the case of closed-loop vertical borehole designs, the thermal barrier presented by a loop can be significant. The high-density polyethylene (HDPE) pipe loop contacts the bore wall via a thermal grout. For example, a 7–8 °C temperature difference between the water circulating in the loop and the bore wall reduces the heat pump operating performance. In general, the loop legs are not constrained within the bore cross section, and during installation, the legs can twist and contact one another over a substantial portion of the bore length, thereby degrading performance. Typically, the loop is sealed to the bore wall with a bentonite grout mixture, of which the hydraulic and thermal properties are very important to insure optimal performance of the installation. The use of a thermally enhanced grout, while advantageous to heat transfer between the loop and the bore wall, presents its own challenges. For example, the addition of sand to enhance conduction also increases thermal shunting between the legs of the loop and presents challenges in uniform mixing and deployment through a tremie pipe.

Acknowledging these challenges, a study was undertaken to understand the potential benefit of multi-loop and custom extrusion designs which offer greater surface area for heat transfer, while locating the pipe proximal to the bore wall. This borehole design strategy could reduce the capital cost of the ground heat exchanger by reducing the amount of drilling required to serve a load. The contractor or design engineer has the option of either reducing the amount of drilling, while achieving the same end-of-season temperature as a standard loop, or retaining the same well depth and achieving better heat pump seasonal performance. The first approach reduces capital cost of the loop field but retains the same annual savings, while the second does nothing to lower capital cost but rather improves annual utility savings. For example, the building owner might be more interested in lowering the loop field installation cost by 20 % and accepting less annual operating savings. Another strategic advantage might derive from less land utilization and associated mess.

The purpose of this publication is to provide an analytical basis for calculating borehole thermal resistance (BTR) for the case of multiple loops and custom extrusions. The desire is to incorporate the borehole model into existing well field sizing software packages, such as GLD, GLHEPRO, and system modeling software, such as TRNSYS (Klein et al. 1996; Shonder & Hughes1997) and HyGCHP, on the market. To accomplish this, the thermal analysis of a single loop (Koenig and Helmke 2014) was chosen as the starting point, acknowledging the contributions of others and developing toward a practical thermal model of a borehole with multiple parallel loops.

Building on the quasi-steady-state modeling (Eskilson 1988) for vertical ground heat exchange, there have been a numerous publications on borehole resistance for a single loop. Previous studies employed finite element numerical techniques (Sharqawy2009; Sagia 2012; Liao 2012). While these are rigorous, they have also proven to be cumbersome to construct. Others have proposed analytical solutions (Hellström 1991; Gu 1998; Philippe 2010) that are readily applied but are restricted to a limited number of pipe geometries. While these studies support the understanding that increasing the grout thermal conductivity and locating the pipe legs proximal to the bore wall enhances borehole performance, it was left to Redmund (1999) and Beier (2012) to actually demonstrate this fact in field tests. A finite element model for the case of two loops in a borehole was published (Diersch 2010), following the work of Al-Koury et al. (2005). Finally, a recent comparative study of helical pipe and triple U-tube loop designs in foundation pile (Zarrella et al. 2013) is included here, as pertinent to this study.

1.1 Borehole thermal resistance for a single loop

A cross-sectional model of the borehole is shown in Fig. 1. The BTR includes contributions from convection inside the pipe (downcomer and upcomer legs), through the pipe walls and through the media that fills the space between the pipes to the surrounding bore wall. The equivalent resistor network model is given in Fig. 2, where the water temperatures inside the pipe loop legs are designated T_1 and T_2, and the surrounding wellbore surface T_b. To simplify the analysis, symmetry is assumed in which the pipe spacing is equidistant from the center line of the bore. The resistance for each pipe, R_p, is composed of convection at the inside wall and conduction through the pipe wall. The shunt resistance, R_s, addresses heat transfer between the pipes through the media, while R_f defines the grout filler resistance between pipe surfaces to the surrounding bore wall.

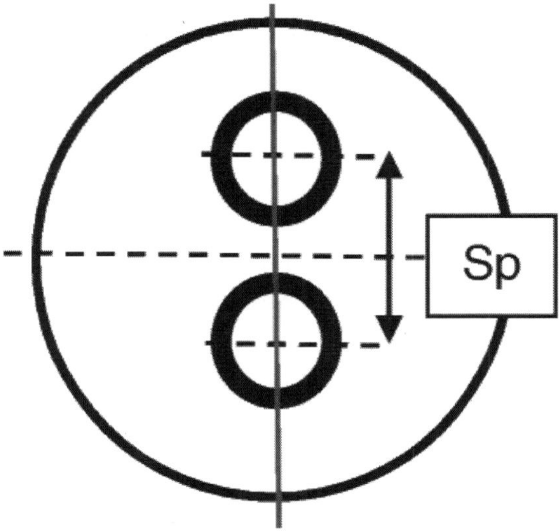

Figure 1. Conduction heat transfer from a pipe loop to the surrounding bore wall

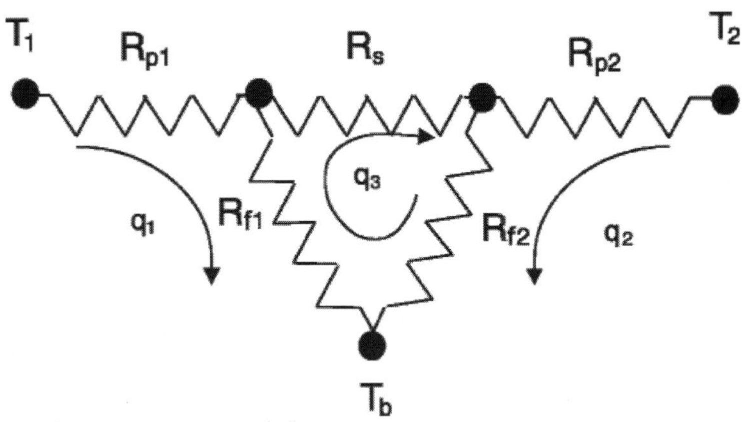

Figure 2. Resistance network model representing well bore heat transfer.

The temperatures, in general, can all be considered different, such that $T_1 > T_2 > T_b$ for the cooling season or $T_b > T_2 > T_1$ for the heating season.

The three equations that describe the resistive network are as follows:

$$(T_1 - T_b) - (q_1 - q_3) R_f - q_1 R_p = 0 \quad (1)$$

$$(T_2 - T_b) - (q_2 + q_3) R_f - q_2 R_p = 0 \quad (2)$$

$$(T_1 - T_2) - q_3 R_s - (q_1 - q_2) R_p = 0, \quad (3)$$

where q is the heating rate per unit length. The total heat transfer between the pipe loop and the bore wall is:

$$q_1 + q_2 = q_t = \Delta T_{avg} / \left[(R_p + R_f)/2 \right], \text{ where}$$
$$\Delta T_{avg} {}^{\circ} (T_{avg} - T_b) \text{ and } T_{avg} = (T_1 + T_2)/2. \quad (4)$$

The total borehole thermal resistance is defined as $(R_p + R_f)/2$.

The three equations can be solved simultaneously for the heating rates using matrix methods:

$\mathbf{qR} = \mathbf{T}$, defined by the vectors $\mathbf{q} = \{q_1, q_2, q_3\}$ and $\mathbf{T} = \{\Delta T_{1b}, \Delta T_{2b}, \Delta T_{12}\}$, where ΔT is the temperature difference designated by the subscripts between the pipes and bore wall. The resistance matrix, **R**, is:

$$\mathbf{R} = \begin{vmatrix} (R_p + R_f) & 0 & (-R_f) \\ 0 & (R_p + R_f) & R_f \\ R_p & (-R_p) & R_s \end{vmatrix} \quad (5)$$

developed from Eqs. (1)–(3). The heating rates, q_i, are found by taking the product of the inverse matrix, $\mathbf{R}^{-1}$, with the temperature column vector. The solution requires an iterative approach, where starting values for the ΔTs are refined by operating within the simulation of the borehole and surrounding rock thermal storage model in time. The temporal effects of borehole heat capacity are minor (quasi-steady-state), insuring that little time (approximately a day) is required to elapse until the ΔTs stabilize. Depending on the loop design basis (m/W), the ΔT_{12} between the loop legs can be considered to be fixed, e.g., −3.3 or 5.6 °C for the heating and cooling seasons, respectively. For the parameters listed in Table 1, the matrix solution is found to be $q_i = \{-24.7, -8.03, -4.69\}$ and $\Delta T = \{-7.3, -3.96, -3.3\}$, respectively, for a loop design based on $(q_1 + q_2) = -32.73$ W/m for the heating season.

Table 1. Parameters chosen for this study

Parameter	Symbol	Size/units
Bore diameter	D	152 mm (6 in.)
Pipe size	d	42 mm (1 1/4 in.)
Pipe c-c spacing	S_p	64 mm (2 1/2 in.)
Pipe wall (HDPE)	DR	11
Pipe thermal conductivity	k_p	0.389 W/m°C (0.225 BtuH/ft°F)
Pipe thermal resistance	R_p	0.089 m°C/W (0.154 ft°F/BtuH)
Grout thermal conductivity	k_g	0.78 W/m°C (0.45 BtuH/ft°F)

The relative amount of heat shunted between pipe is calculated from the ratio $q_3/(q_1 + q_2)$. This turns out to be an important and useful number, because it defines what fraction of the pipe surface area is effective in heat transport to/from the bore wall. This will be apparent shortly in the shape factor analysis of individual pipe heat transfer.

The values for the pipe resistance, R_p, the pipe to bore wall resistance, R_f, and the shunt resistance, R_s, are developed here assuming steady-state conditions. The pipe resistance is composed of series resistances due to convection at the interior wall from water heat transfer and pipe wall contribution. These can be calculated as follows:

$$R_p = 1/(\pi d_i h_{cv}) + \mathrm{LN}(d_o/d_i)/(2\pi k_p), \tag{6}$$

where the d subscripts (o, i) reflect outside and inside pipe diameters, and h_{cv} is the convection coefficient for water flowing inside the pipe, and k_p, the pipe material thermal conductivity. Note that $d_o/d_i = 1/(1 - 2/\mathrm{DR})$, where DR is the dimension ratio $= d_o/t$, where t is the thickness of the pipe wall.

Similarly, the thermal shunt resistance, R_s can be modeled as pipe to pipe conduction shape factor in an infinite medium (Holman1981), although the borehole is finite.

$$R_s = \cosh^{-1}(x_{pp})/(2\pi k_g), \text{ where } x_{pp} = [2(Sp/d)^2 - 1] \tag{7}$$

Defining dimensionless quantities: $d_r = d/D$ and $S_r = S_p/d$, the parameter, x_{pp}, can be written as:

$$x_{pp} = [2S_r^2 - 1]. \tag{7A}$$

For example, Table 1 lists select parametric values in a typical geo-exchange application:

The grout filler resistance between pipe and bore wall is more complicated. Referring to Fig. 2, the shape factor for heat transfer for a single offset pipe to its surrounding cylinder was compiled by Holman (1981):

$$R_{fs} = \cosh^{-1}(x_f)/(2\pi k_g), \text{ where } x_f = (D^2 + d^2 - Sp^2)/(2Dd). \tag{8}$$

where D is the bore diameter, d the pipe exterior diameter, S_p the center to center (c-c) spacing between pipes, and k_g the grout thermal conductivity. Again, this can be defined in dimensionless form as:

$$x_f = ½\,[d_r(1 - S_r^2) + 1/d_r]. \tag{8A}$$

For the selected parameters in Table 1, for example, the pipe to bore resistance is 0.219 m°C/W. Equation (8), however, is based on a single off-centered pipe conducting to an unobstructed surrounding wall; therefore, one might anticipate that the presence of a second pipe (leg of the loop) would contribute a similar measure, so that the total would simply be half of the single pipe contribution, as suggested by the resistance mesh model given earlier. Unfortunately, this is not quite right in that there is conductive interference due to the presence of the pipe representing the second leg of the loop.

This interference or obstruction is precisely addressed by including the factor, f, defined as:

$$f = 1 - q_3/(q_1 + q_2), \tag{9}$$

so that $R_f = R_{fs}/f$ in the denominator of the pipe to bore resistance, R_{fs}, to correct the thermal influence of each pipe. The *total* borehole resistance, in this case, is 0.172 m°C/W. This is composed of parallel contributions of each pipe individually: $0.089 + 0.219/(1 - 0.143) = 0.345$ m°C/W, including the pipe and the grout. For the loop, this value is halved to yield the total borehole resistance stated. Note that the pipe legs are at different temperatures, resulting in the thermal shunt of 14.3 %.

A closed-form solution for the pipe loop to bore resistance, R_{pb}, develops from solving:

$$\mathrm{RI} = 1/R_f = 2\pi k_g f / \cosh^{-1}(x_f), \tag{10}$$

where $f = 1 - \Delta T/(q_t R_3)$, ΔT is the loop delta-T, q_t the total heating rate, and $R_3 = (2R_p + R_s) + (R_p R_s)/R_f$.

The latter results from solving the matrix equations for q_3 in terms of the heating rate difference $(q_1 - q_2)$.

The quadratic solution is:

$$\mathrm{RI} = -\tfrac{1}{2}(a/b - 1/R_f)\left\{1 - \mathrm{SQRT}\left[1 - 4(c/b - a/b\, R_x)/(a/b - 1/R_{fs})^2\right]\right\}, \tag{10A}$$

where $a = (2R_p + R_s)$, $b = R_p R_s$, $c = \Delta T/(q_t R_3)$, R_{fs} and R_3 as defined in (8) and (10), respectively.

The final borehole resistance (BTR) for a single loop has a closed-form solution that is summarized by:

$$R_b = (R_p + R_f)/2, \text{ where } R_f = 1/\mathrm{RI}. \tag{11}$$

The benefit of the closed-form quadratic solution is that it avoids the need for the **T** information, which requires the full bore simulation, by providing a steady-state value for R_f straightaway. The two approaches give the same result, with the matrix solution offering additional information on **q**.

1.2 Comparison to others

In prior work (Koenig and Helmke 2014), which provided a comparison to other reported models, a similar approach was employed to assess single-loop BTR, with the difference being in the use of a geometric obstruction factor, $f(\beta)$. The comparison, while quite good for larger separations, was shy of reported results for cases in which the two pipes were close to one another. As Fig. 3 demonstrates in a comparison of obstruction factors, the discrepancy is confined to the use of a geometric obstruction factor for pipe separations within 35 % of the bore diameter.

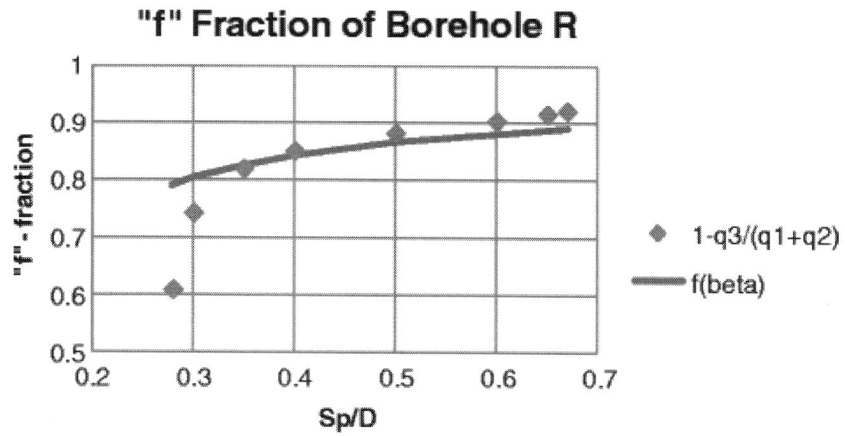

Figure 3. Comparison of obscuration factors for this model (*blue*) vs. geometric (*red*)

A comparison of this improved model to the results of Sharqawy, shown in Fig. 4, clearly shows a very good match over the full extent of the pipe separation study. Note, at $S_p/D \approx 0.28$, the bore resistance experiences a jump as the pipe legs actually contact. This result lends credence to the model proposed here.

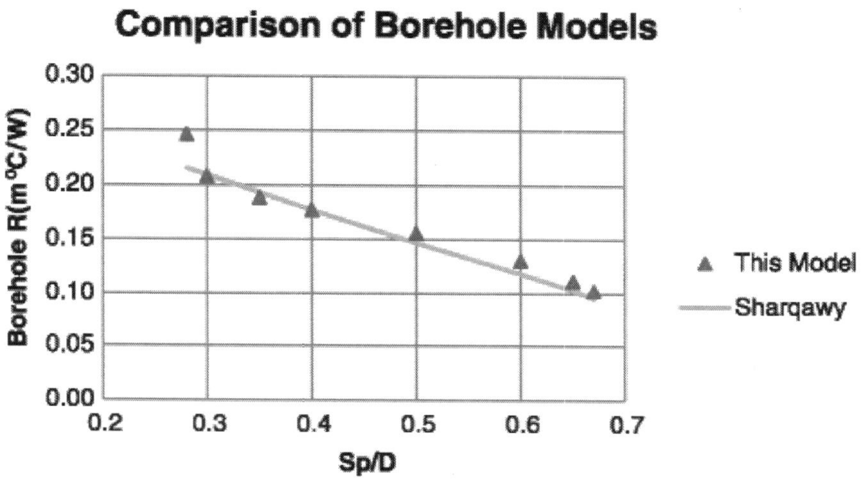

Figure 4. Comparison of borehole resistance models

Figure 5 demonstrates the benefit of employing higher conductivity grout (reported in W/m°C) as well as increased pipe separation in lowering borehole resistance. Even though thermal shunting increases with higher grout conductivity, the impact of greater pipe conduction to the bore wall tends to dominate the response. In all cases, however, there is a 25 % net BTR benefit to locating the pipe proximal to the bore wall.

2. METHODS

2.1 Extension to multi-loop borehole geometries

In principle, adding additional loop(s) to a borehole should decrease borehole resistance by increasing the available pipe surface area for heat transfer. It should be noted, however, that the heat transfer effectiveness of an individual loop diminishes with each additional loop. This is because the heat transfer surface area of an individual loop for exchange with the bore wall is occulted by the presence of additional loops.

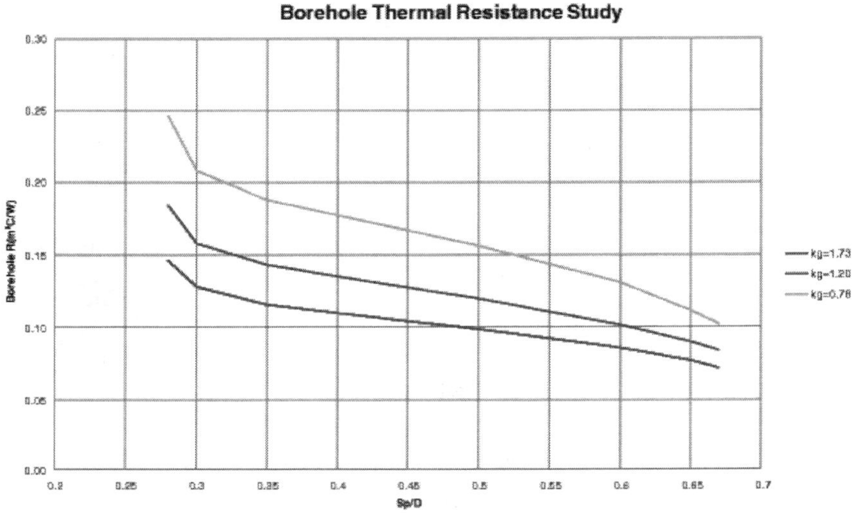

Figure 5. Borehole resistance study as a function of grout thermal conductivity

The method delineated here extends the single-loop BTR model by first defining the boundary in the borehole cross section that separates downcomer and upcomer pipe legs of a common temperature. Pipe loops are assumed to be connected in parallel. The thermal interaction between individual pipes across the boundary is treated independently and, without assigning legs of a common loop, using the single loop methodology presented previously. The total borehole resistance is assembled by summing the individual pipe-pipe resistances of a common spacing weighted by their geometric configuration factor. The configuration factor is developed by noting the number of occurrences of specific pipe-pipe geometries within the borehole. An example is offered to help clarify the procedure.

Starting with the two-loop borehole, shown in Fig. 8 leftmost diagram, the values of R_f are calculated for individual pipe on one side of the boundary line to the two pipes on the other side (diagonal and adjacent configurations). The resistances, R_p, R_{pp}, and R_f, are calculated for each configuration. The final (Fig. 6) BTR is composed of the contributions of each configuration. The example in Table 2 for $D = 152$ mm, $d = 33$ mm, and $k_g = 0.78$ W/m°C shows the assemblage of the final BTR value from equal contributions of each configuration.

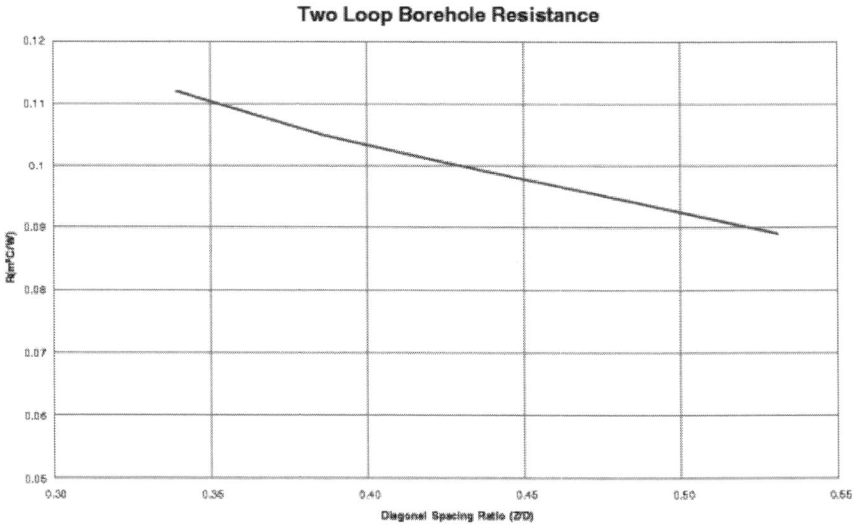

Figure 6. Borehole resistance (BTR) for 2U design as a function of relative loop leg spacing (Z/D)

Table 2. Two-loop borehole resistance model results

N_{100p} =	2		
Case 1.	(mm)	f	R_b
Z =	63.5	0.878	0.196
$Z/\sqrt{2}$ =	44.9	0.828	0.205
		Avg/2N_L =	0.100

To conform to the borehole, the ordered pipe planform geometries become square (2U), hexagonal (3U), and octagonal (4U), as indicated in Fig. 8. Pragmatically, in order to accommodate these configurations, the outside diameter of the assembly must remain within a maximum envelope, e.g., 114 mm in a 152-mm bore, or 75 % of the opening. To accomplish this, pipe sizes must shrink from 42 mm (1 1/4 in.) to 33 mm (1 in.) down to 27 mm (3/4 in.) as the number of loops increases. The constraints on allowable geometry impose restrictions on the range of pipe separations that are feasible. Moreover, each of these multi-loop designs requires an assembly to affix the pipe to the desired geometry.

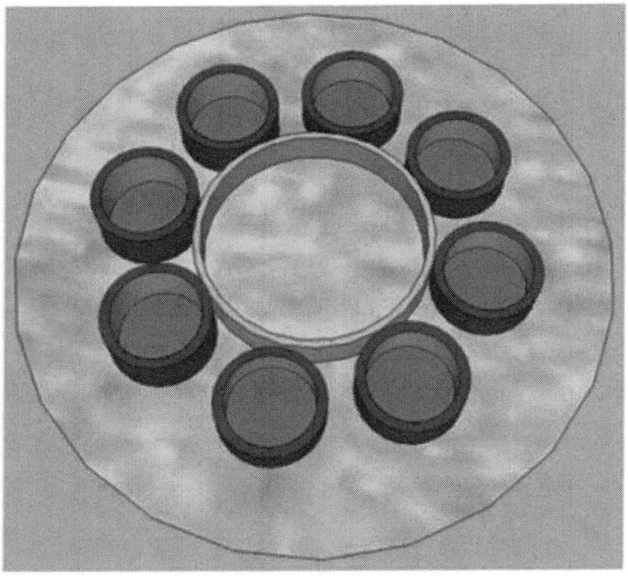

Figure 7. Four-loop borehole configuration

As an example, Fig. 7 shows four loops (eight pipes) arranged evenly around a hollow central form of diameter, d_{cyl}.

The maximum (even) number of pipes $N_x = 2N_L$ that fit around the form is limited by:

$$d \leq \left(d_{\text{cyl}} + d\right) \sin(\pi/N_x), \tag{12}$$

where d_{cyl} is the outside diameter of the form. In addition, there is a practical constraint based on the maximum diameter of the loop assembly that the borehole can comfortably accept,

$$d_{cyl} + 2d \leq 0.75D \tag{12A}$$

where 75 % is deemed a reasonable limit. Combining (12) and (12A) leads to the condition:

$$[0.75(D/d) - 1] \sin(\pi/N_x) \geq 1 \tag{12B}$$

So, for example, with a 152-mm (6 in.) bore and a 27-mm (3/4 in.) pipe size, the maximum number of pipes that fit the constraint is ten or, equivalently, five loops. The outside diameter (OD) of this assembly is then ($d_{cyl} + 2d$), or, in this example, 114 mm (4.5 in.). Other sizes are possibly limited only by the assembly size impact on insertion in the borehole and the ability to insert a grout "tremie" pipe inside the ID of the form, or, ideally, use the form itself as the grout pipe.

To extend the previous model to the case of multi-pipe borehole geometries, under the assumption that the form itself offers negligible thermal isolation, one needs to simply specify the number of pipe loops, the diagonal pipe spacing, and then calculate the pipe-pipe shunt resistance for each configuration. The latter is based on the pipe-pipe spacing for each configuration, as shown in Fig. 8.

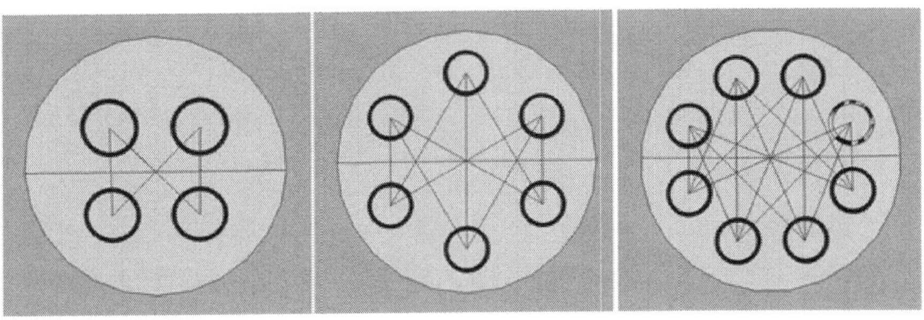

Figure 8. Compilation of thermal separations between pipes for two-, three-, and four-loop assemblies

The horizontal line delineates pipe of a common temperature (downcomer vs. upcomer legs). Table 3 summarizes the spacing and number of configurations for each loop geometry, where Z is the fixed diagonal c-c spacing. From this table, the shunt spacing for each separate configuration is identified and the shunt resistance, R_s, can be computed for each configuration separately. This value, then, is inserted into the borehole resistance matrix to derive the obstruction factor, f.

Table 3. Multi-loop pipe-pipe spacing for shunt analysis

N loops	S_p	Configuration occurrences	Configuration fraction (CF)
1	Z	1	1
2	Z	2	1/2
	$Z/\sqrt{2}$	2	1/2
3	Z	3	1/3
	0.866 Z	4	4/9
	$Z/2$	2	2/9
4	Z	4	1/4
	0.9239 Z	6	3/8
	0.707 Z	4	1/4
	0.3827 Z	2	1/8

The final borehole resistance for the multi-loop assembly is calculated from:

$$\mathrm{BTR} = \Sigma\left[R_{\mathrm{bi}}(\mathrm{CF})_i\right]/N_{\mathrm{L}} \quad (13)$$

where R_b is the borehole resistance for a single (Z) configuration, multiplied by CF, its configuration fraction, all summed over the number of configurations from $i = 1$ to N_L.

One of the advantages of additional loops is seen in the "N_L" factor that appears in the denominator which decreases borehole resistance as loops are added. An additional benefit is that the pipe wall thickness, assuming a fixed DR for pressure rating, decreases as the pipe gets smaller to accommodate the loops.

3. RESULTS

Using the method outlined in this paper, a compilation of multi-loop borehole resistance is provided in Fig. 9 for specific loop designs and values of the pipe spacing ratio, *d*/*D*. The reduction in borehole resistance as additional loops are added is dramatic. This is especially noticeable with the addition of a second loop.

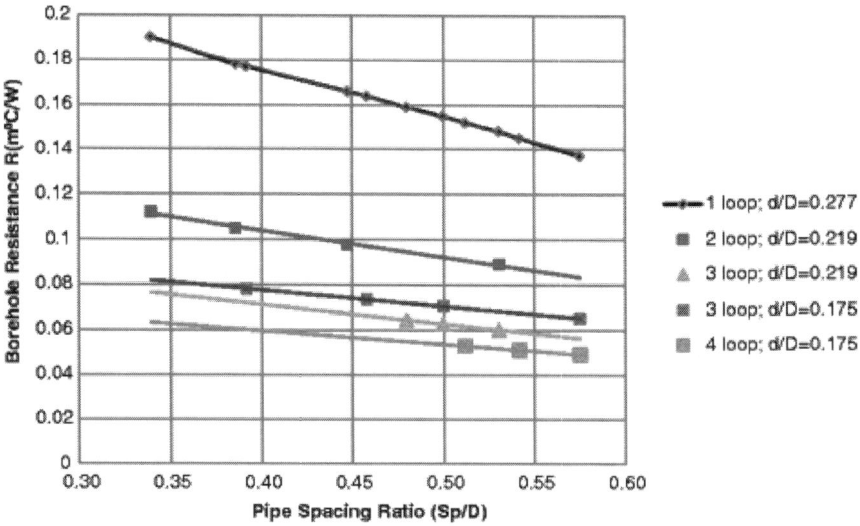

Figure 9. BTR at $k_g = 0.78$ W/m°C as a function of relative pipe spacing and loop designs

By constraining the OD of the multi-pipe loop assembly to 75 % of the bore diameter, the study considered the benefit of additional loops. Fig. 10 shows that the borehole resistance decreases dramatically and was found to be inversely proportional to the number of loops to the power of 0.86.

Figure 10 compares BTR values of multiple loop designs reported by specific loop manufacturers to the predictions of this model. In addition to the single-loop match (Fig. 4), there are two additional points of validation:

- From thermal measurements by Bowman Geothermal of REHAU 1-in. double-loop product in side-by-side installations, they report a BTR of 0.11 (PEX_2), a 40 % reduction from single loop, which, after adjustment for HDPE, matches the model with an RMS error of 1.5 %.

- Agreenability (2014) report a BTR of 0.049 m°C/W in an application of their Twister™ four-loop 3/4-in. pipe technology, which compares favorably to predictions exhibited in Fig. 9.

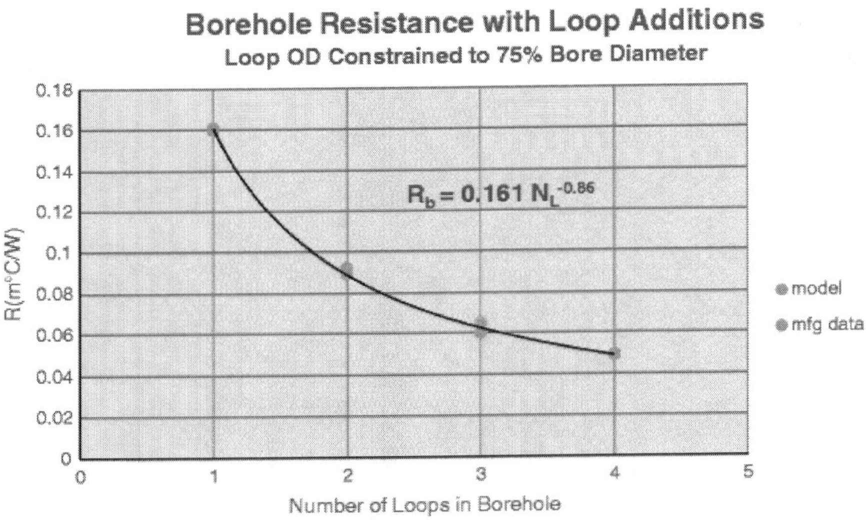

Figure 10. BTR decrease as loops are added to the borehole

At some point with addition of multiple loops, the borehole resistance model effectively transitions to a central pipe equivalent diameter, d_{eq}, with a grout gap defined by:

$$R_b = R_p/2N_L + \mathrm{LN}(D/d_{eq})/(2\pi k_g), \tag{14}$$

where $(Z+d) \leq d_{eq} \leq (2N_L - 2)d$ for $N_L = 2$, 3, or $(2N_L - 1)d$ for $N_L = 2$. The equivalent diameter is the wetted pipe perimeter limited between the smooth OD of the assembly and the exposed scalloped perimeter of the individual pipe. In this case, the pipe wall between pipes in the assembly becomes adiabatic, except for the end pipe that are at different temperatures.

If the assembly was designed such that the adjacent pipe in the assembly was at different temperatures, then, of course, this would diminish performance because of thermal shunting. Rather, if the assembly could be arranged to segregate pipe of the same temperature (downcomers and upcomers), as suggested in Fig. 11a, then, this might indeed improve performance. An alternate loop design, shown in Fig. 11b, eliminates the intermediate adiabatic walls between pipe in Fig. 11a, in favor of a custom kidney extruded shape. This design by ARB Geowell (2014) eliminates the need for an assembly of smaller pipe in favor of coiled pipe halves that must be joined with a molded transition from kidney pipe to 1 1/2-in. round pipe for connections.

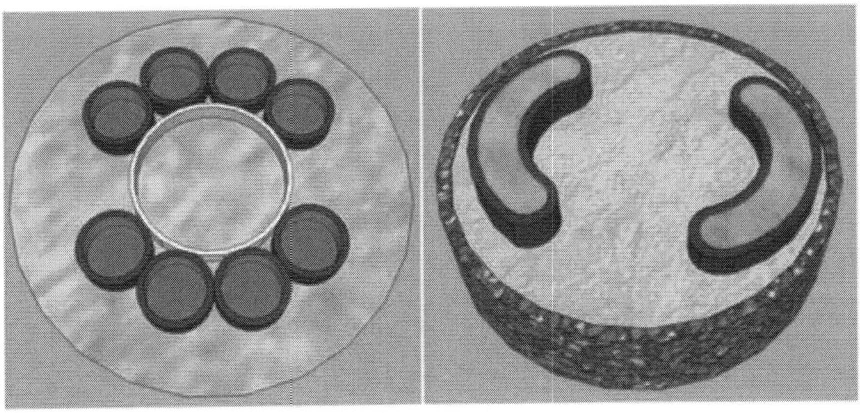

Figure 11. **a** Segregated pipe assembly. **b** Extruded kidney pipe deployment

The utility of the proposed borehole thermal model is such that it can be easily extended from simple loops to more complex multi-pipe and custom shape loop configurations. The model addresses only the thermal resistance inside the borehole (BTR), and not the diffusion component into the surrounding formation. The benefit is that it can be incorporated easily into existing software, such as GLHEPRO, to handle multi-loop borehole configurations, avoiding the details of constructing custom numerical models through FEFLOW.

The model produces reasonable results that support the data published by others. The extension of the model to multiple pipe loops in a single borehole is accomplished by (1) defining the mid-plane boundary between downcomer and upcomer pipe, (2) calculating the borehole resistance for each pipe-pipe separation across the boundary, (3) multiplying by the relative number of occurrences of each configuration to yield an average borehole resistance for a representative loop, and finally (4) dividing by the number of loops. Since the loop to bore wall temperature difference is controlled by the product, $q(R_b)$, it is noteworthy that it makes no difference whether this is treated as a single loop, in which case $(q/N_L)R_b$, or the full multi-loop design, $q(R_b/N_L)$.

The benefit and practicality of implementing multiple loop borehole designs to improve performance and/or reduce installation cost is arguable. The use of a thermally enhanced grout with a single loop appears to lower borehole resistance by 30 % over neat bentonite grout at standard U-bend separations, despite higher thermal shunting between the legs of the loops. But, unless the pipe legs can be guaranteed to be separated and not twisted and contacting over a substantial length of the bore, this 30 % benefit may never be realized in practice. Physical separation, such as the use of Geo-Clips, is at least as beneficial as thermally enhanced grout. With this in mind, the obvious benefit of multiple loops in reducing borehole resistance is increased surface area for heat transfer and thinner HDPE walls that result from employing smaller diameter pipe loops. Another advantage to multiple loops is in lowering the pressure drop of the ground heat exchanger, which can be as much as 35 %. This benefit develops from dividing the flow into multiple pipe loops of a given cross section as well as shortening the length. In operation, this can save considerable pump energy over the year.

With that said, the question remains whether the assembly of pipe loops can be cost-effective with the benefit of less required drilling. On paper, the savings in drilling more than offsets the extra cost of multiple pipe assembly, which ultimately means that plastic is less expensive than the lifecycle operation of the drill rig. Certainly, there are excellent targets for commercialization of new loop technology, such as commercial applications where space for the loop field is always an issue, and the opportunity to repurpose abandoned water wells that could serve residential customers without extending the depth of the well. Ultimately, driller pricing and competition will prove the winner.

4. CONCLUSIONS

A borehole thermal model for vertical loops was developed from a modified shape factor analysis. An interference factor, f, defined by $(1 - q_{shunt}/q)$, the relative size of the thermal shunt between the loop segments, was employed as the modification required to properly model borehole thermal resistance. The model predictions were compared to reported results and found to match quite nicely over a range of pipe sizes and spacing. Following this success, the model was extended to multi-pipe loop geometries consisting of two-, three-, and four-loop assemblies in a single borehole and a custom kidney extrusion. The borehole resistances were calculated for each of these configurations and presented as a compilation of predicted performance. These predictions were compared to the available data reported by manufacturers of multi-pipe loop assemblies and found to give reasonable results.

COMPETING INTERESTS

The author declares that he has no competing interests, although ARB Geowell owns the patent for K-Pipe, a kidney pipe custom extrusion. The performance benefit of this pipe was treated objectively within the context of multiple pipe loops. The author has no financial commitment from the company to support the publication of this work nor derives a benefit from its publication.

REFERENCES

1. Al-Koury R, Bonnier PG, Brinkgreve RB (2005) Efficient finite element formulation for geothermal heating systems. Part I: steady-state. Int J Numerical Methods Eng 63(7):933–1013, Pt. II: Transient, IJNME **67** (2006) 5, 725–745

2. Agreenability, J R (2014) TWISTER high efficiency geothermal heat exchanger. IGSHPA Conference, Baltimore, MD

3. ARB Geowell, Koenig A (2014) A new HDPE pipe loop for single well residential opportunity. Paper presented at the PGWA winter conference, Grantsville, PA, 23–24 January 2014

4. Beier RA, Ewbank GN (2012) In-situ thermal response test interpretations: OG and E ground source heat exchange study. IGSHPA, OKSU

5. Crecraft H, Mahlmann J, Rehau in-situ thermal performance testing Bowman Geothermal, LLC, Chantilly, VA: http://www.rehau.com/download/866550/borehole-thermal-conductivity-report.pdf

6. Diersch HJ, Bauer D, Heidemann W, Rühaak W, Schätzl (2010) Finite element formulation for borehole heat exchangers in modeling geothermal heating systems. FEFLOW, DHI-WASY software, White Papers Volume V. 5:5–96.

7. Eskilson P, Claesson J (1988) Simulation model for thermally interacting heat extraction boreholes. Numerical Heat Transfer 13(2):149–165

8. GLD: www.seventhwave.org/hygchp

9. GLHEPRO: https://hvac.okstate.edu/glhepro/overview

10. Gu Y, O'Neal D (1998) Development of an equivalent diameter expression for vertical U-tubes used in ground-coupled heat pumps. ASHRASE Trans 104(2):9

11. Hellström G (1991) Ground heat storage, thermal analyses of duct storage systems, doctoral thesis. Department of Mathematical Physics, University of Lund, Lund, Sweden

12. Holman JP (1981) Heat transfer. McGraw-Hill, New York, pp 68–71

13. HyGCHP: www.ecw.org/hygchp

14. Koenig A, Helmke M (2014) Development of a thermal resistance model to evaluate wellbore heat exchange efficiency. Int J Energy Environ 5(3):297–304

15. Liao Q, Zhou C, Cui W, Jen T (2012) Effective borehole thermal resistance of a single U-tube ground heat exchanger. Numerical Heat Transf 62:197–210

16. Phillippe M, Bernier M, Marchio D (2010) Sizing calculation spreadsheet vertical geothermal borefields. ASHRAE J 7:20–28

17. Redmund CP (1999) Borehole thermal resistance: laboratory and field studies. ASHRAE Trans 105:439–445

18. Sagia Z, Stegou A, Rakopoulos C (2012) Borehole resistance and heat conduction around vertical ground heat exchangers. The Open Chem Eng J 6:32–40

19. Sharqawy MH, Mokheimer EM, Hassan MB (2009) Effective pipe-to-borehole thermal resistance for vertical ground heat exchangers. Geothermics 38:271–277

20. TRNSYS: sel.me.wisc.edu/trnsys/downloads/trnsedapps/gchpdemo/overview.htm; Klein, S.A., et al., TRNSYS Manual, "A Transient Simulation Program", Solar Energy Laboratory, University of Wisconsin, Version 14.2, for Windows, Sep 1996.

21. Shonder, John and Hughes (1997) Patrick from ORNL (Oak Ridge National Laboratory). Richard Stockton College.

22. Zarrella A, DeCarli M, Galgaro A (2013) Thermal performance of two types of energy foundation pile: helical pipe and triple U-tube. Appl Therm Eng 61(2):301–310

CHAPTER 12

3-D finite element analysis of roller-expanded heat exchanger tubes in over-enlarged tubesheet holes

N. Merah[1], A. Al-Aboodi[2], A. N. Shuaib[1] and Y. Al-Nassar[1]

[1]Mechanical Engineering Department,King Fahd University of Petroleum and Minerals, Dhahran 31261, Saudi Arabia

[2]Mechanical Engineering Department, Qassim University, PO Box 6677, Buraydah 51452, Saudi Arabia

ABSTRACT

Repeated de-stubbing and re-tubing of heat exchangers in petrochemical plants during their useful lifetime may result in over-enlarged tubesheet holes with overtolerances that exceed those prescribed by Tubular Exchanger Manufacturing Association (TEMA) standards (1988). Roller expansion of tubes in these over-enlarged holes may result in tube thinning and weakening of the joint due to a decrease of interfacial pressure between the tube and tubesheet. In the present work, a 3-D finite element (FE) model of a tube-tubesheet joint was used to determine displacement and stress distributions along the axial direction of roller expanded tube-tubesheet joint and to evaluate the combined effects of large initial clearance and strain hardening of tube material on interfacial pressure and tube deformation. The results obtained from the present model are compared to those of axisymmetric FE analysis and to the experimental results. Both axisymmetric and 3-D models are found to yield comparable trends showing that for elastic perfectly plastic tube material the residual contact pressure remains constant well above the prescribed TEMA maximum overtolerance values. In addition, both models show that for strain hardening tube materials the interfacial pressure increases with increasing clearance. An appreciable difference is observed at high overtolerances where the 3-D model predicts cut-off clearances (clearance at which the interfacial pressure starts to drastically drop) which are about 30% lower than those predicted by the axisymmetric models. The tube inner surface deformation and pull out forces estimated from 3-D results compares very well with those obtained from the experimental tests.

KEYWORDS

3-D Finite element, Tube-tubesheet joint rolling, Clearance effects, Strain hardening, Overtolerances

1. INTRODUCTION

The frequent process of de-stubbing and re-tubing operations performed during the life of the heat exchangers in petrochemical and energy producing plants results in the over enlargement of the tubesheet hole to levels that may exceed the levels prescribed by industrial standards (Standard of the Tubular Exchanger Manufacturer Association 'TEMA' 1988). When new tubes are installed, they are often expanded using the rolling process into the enlarged tubesheet holes; this may result in a deterioration in the strength of tube–tubesheet joint having large initial clearance as well as in the strain hardening and thinning of the expanded tubes and their surrounding ligaments.

Simulation of roller expanding of tubes into tubesheet by the finite element method is associated with a number of difficulties such as roller kinematics and the absence of loading axisymmetry. In fact, the mechanical rolling process is a good example of periodic symmetry with respect to the load application. A number of researchers have shown that the complexity of the tube–tubesheet roller expansion can be simplified by reducing it to an axisymmetric quasi-static problem (Aufaure 1987; Merah et al. 2010, 2009; Updike et al. 1992; Merah 2008; Al-Aboodi et al. 2009; Cizelj and Mavko 1995; Williams 1996, 2003). Displacements are applied in steps up to the required expansion and then decreased down to zero such that the tube–tubesheet contact ensures the strength of the connection joint after transmitting a large radial stress to tube's outer surface. Andrieux and Voldoire (1995) developed an axisymmetric FE model to simulate roller expansion using five steps. Cizelj and Mavko (1995) used the axisymmetric model in ABAQUS code to evaluate the residual stresses in one-step rolled tube to tubesheet joint. Williams (1996), used three load steps to complete the final mechanical roll expansion. Because of the absence of loading symmetry other researchers, however have used 3-D finite element models to simulate roller expansion of tube–tubesheet joints (Williams 1997; Metzger et al. 1995). These models which are more realistic provide more in-depth information about the displacement and stress distributions in all directions Williams however, concluded that the 3-D model did not bring any additional information when compared to axisymmetric model (Williams 1997).

To the authors' knowledge the effect of tube–tubesheet initial clearance on the strength of tube/tubesheet joint has not been addressed by 3-D FE analysis of roller expansion. Existing work has relied mainly on axisymmetric models with hydraulic expansion (Allam et al. 1998) or with roller expansion (Al-Aboodi et al.2008, 2009; Cizelj and Mavko 1995; Williams 1996).

In this paper, a 3-D FE model has been utilized in a major commercial finite element analysis code (ANSYS2004) to simulate the roller expansion and study residual deformation and stress distributions to show the effect of initial clearance and tube material on the residual contact pressure between the tube and tubesheet. Initial clearances within and beyond the prescribed TEMA range (Standard of the Tubular

Exchanger Manufacturer Association 'TEMA' 1988) have been considered. Low-to-high tube and tubesheet material strain hardening properties have also been included in the study and their effects discussed herein. The 3-D model results pertaining to the effect of clearance on interfacial pressure are compared with those obtained from the axisymmetric simulation of the same effects and validated using experimental results developed earlier by Shuaib et al. (2003).

1.1 Geometry and 3-D FE Model

The equivalent sleeve diameter determined earlier by Shuaib et al. (2003) for the stabilizer feed/bottom exchanger was implemented in the 3-D model. Figure 1a illustrates the geometry and dimensions of the model of tube–tubesheet joint under study. Figure 1b shows the full 3-D model mesh and Fig. 1c a 2-D view of the model with boundary conditions on the primary side.

Meshing of tube and tubesheet is performed using 3-D structural solid element ANSYS-SOLID 186 which has a quadratic displacement behavior and is well suited for modeling irregular meshes. The element is defined by 20 nodes having three degrees of freedom each: translations in the nodal x, y, and z directions. The element is chosen here because it supports plasticity, hyperelasticity, and large deflections. The model of Fig. 1b has 8,000 elements and 9,920 nodes. The level of refining was tested to establish effective meshed models. It was found that refining the mesh beyond 8,000 elements did not add any significant precision to the solution.

The contact element CONTA174 is used to represent contact and sliding between 3-D "target" surfaces (TARGE170) and a deformable surface, defined by this element. The element is applicable to 3-D structural and coupled field contact analyses. This element is located on the surfaces of 3-D solid elements with mid-side nodes. It has the same geometric characteristics as the solid element face with which it is connected. Contact occurs when the element surface penetrates one of the target segment elements (TARGE170) on a specified target surface. A sensitivity analysis was performed in order to select the real constants values such as those defining tolerance and offset between the contact and target elements (Al-Zayer 2001; Al-Aboodi 2006).

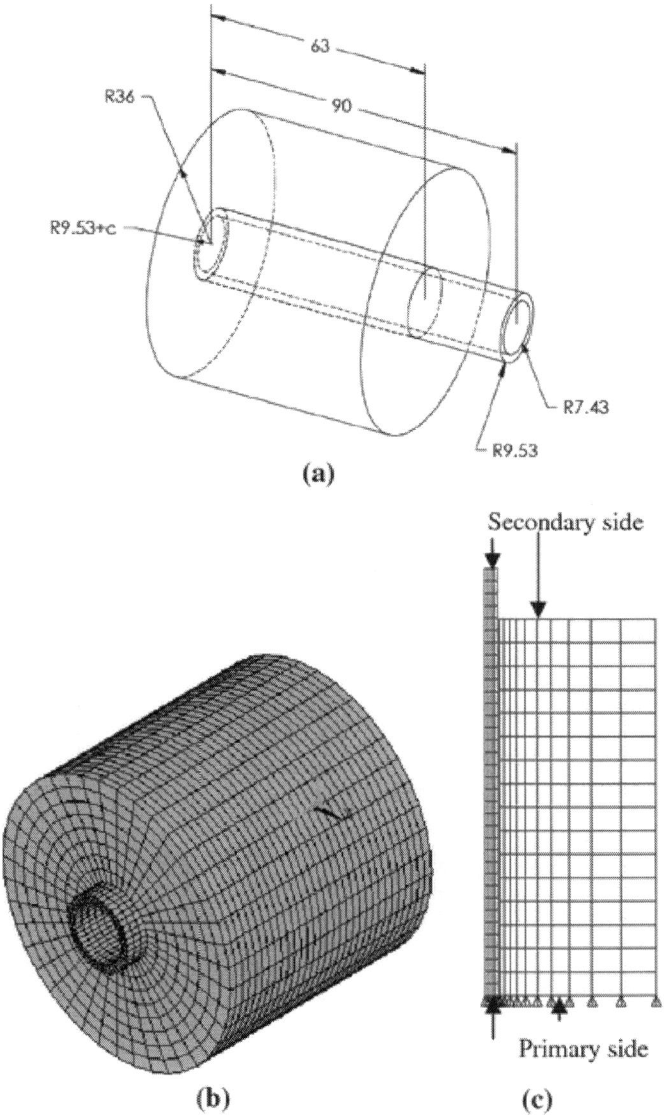

Figure 1. Geometry and FE mesh of the model. **a** Tube and tubesheet assembly with dimensions in millimeters, **b** full 3-D model mesh and **c** 2-D representation showing boundary conditions on primary side

1.2 Joint materials

The elastic–plastic behavior of tube and tubesheet material is represented by bilinear curves. In the experimental work used here for validation, the tube is seamless, cold-drawn, low-carbon steel-type ASTM 179 and the tubesheet material is carbon steel-type ASTM A5 16 G70 with average yield stress of 248 MPa and modulus of elasticity of 207 GPa (Al-Aboodi et al. 2008; Shuaib et al. 2003).

Because of expected large plastic strains, the Bilinear Isotropic Hardening (BISO) option was used in the 3-D FE model. The curve in the plastic region was approximated by a linear relationship. The slope of the approximated line in the plastic region of the true stress–strain diagram defines the tangent modulus of plasticity (E_{tt}). An elastic-perfectly plastic material is that having zero tangent modulus. The average value of E_{tt} for the tube material used in the experimental study was found to be 733 MPa (Al-Aboodi et al. 2008). In order to investigate the effect of tube material strain hardening on the residual stresses, E_{tt} values ranging from 0 to 1.2 GPa were considered (Table 1). These values were chosen to cover the plastic behavior of most of the steel materials used in similar applications. Table 1 contains the input parameters to FE code of the analysis. They consist of the range of clearances (c), percentage wall reductions (%WR), and tube and tubesheet material constants. The friction coefficient between the tube and tubesheet is assumed to be zero (Merah et al. 2003; Al-Zayer 2001).

Table 1. Input parameters for 3-D FE model

Parameter	Value or range of values
Initial clearance, *c*	0–4.5 mm
Wall reduction (%WR)	5%
Tube and tubesheet material elastic modulus, *E*	207 GPa
Tube and tubesheet material yield stress	248 MPa
Poisson's ratio, ν	0.3
Tangent modulus, E_{tt}	0, 0.4, 0.8 and 1.2 GPa

1.3 Simulation of rolling

Rolling is performed by the rigid rollers that stay in contact with the inner surface of the tube being expanded. During expansion the rollers are modeled as rigid lines that undergo outward radial displacement and simultaneous circumferential motion that simulate tube expansion kinematics. According to the industrial rolling practice specified by a major roller expansion equipment manufacturer (Cooper Power Tools2005), the amount of the radial displacement (ur)(ur) of the rigid rollers against the tube inner surface during the tube expansion phase is obtained from the values of the required tube percent wall reduction (%WR = 5%), the tube thickness (t), and initial tube–tubesheet clearance (c = 0–4.5 mm) as follows:

$$u_{\mathrm{r}} = (\%\mathrm{WR})t + c. \tag{1}$$

Once the expansion process reaches this limit of radial deformation, the inner tube surface will be released. The rolling process is simulated by three rigid rollers, 120° apart, moving circumferentially by increments of 6°. Four revolutions are required to reach the maximum radial displacement determined by Eq. 1. The rolling process is described schematically in Fig. 2a. Graphical representation of the displacement profile along the circumference of the inner tube is shown in Fig. 2b. Figure 3 shows that for each roller, it is assumed that a total of 11 nodes are effectively deformed. Five nodes on each side of the roller are partially displaced plus the middle node along the generating line. Equation 2 is used to define the deformation profile yielding maximum displacement at the mid node. It is a sinusoidal function that represents the shape of the inner tube surface displaced by the radial movement of the roller.

$$u_i = u_{\mathrm{s}}\, \mathrm{e}^{-\beta_i} \left(\sin(\beta_i) + \cos(\beta_i)\right) \tag{2}$$

where uiui is the nodal displacement at the ith node, usus is the step displacement and β_i is a normalized nodal factor taking the values 0, 0.25, 0.5, 0.75, 1.0, and 1.25; $\beta_i = 0$ corresponds to the mid point.

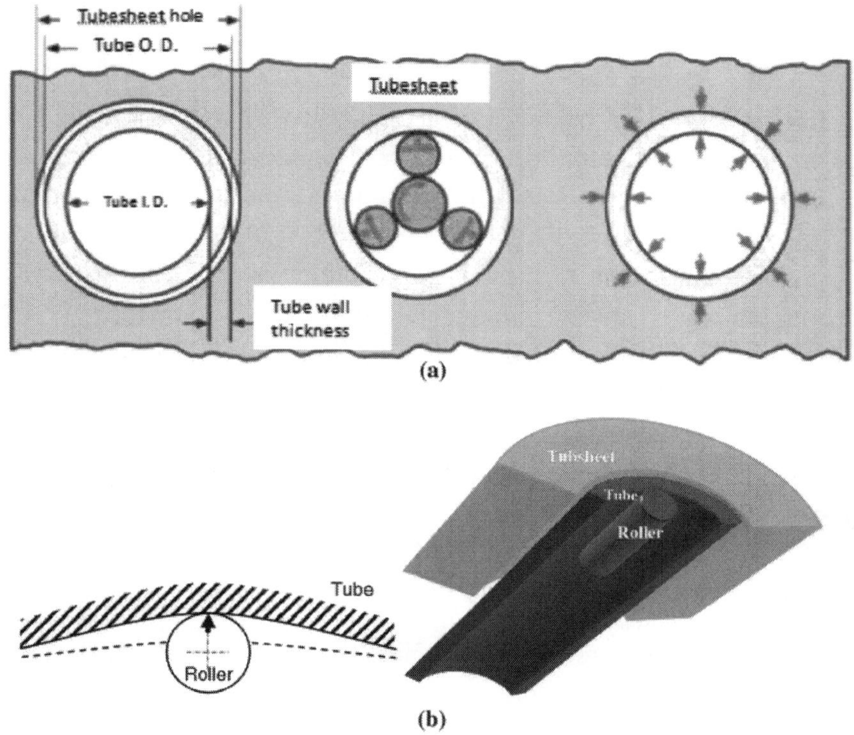

Figure 2. a Schematic representation of the industrial rolling process (Cooper Tools 2005), **b** Graphical representation of rolling process in the present model

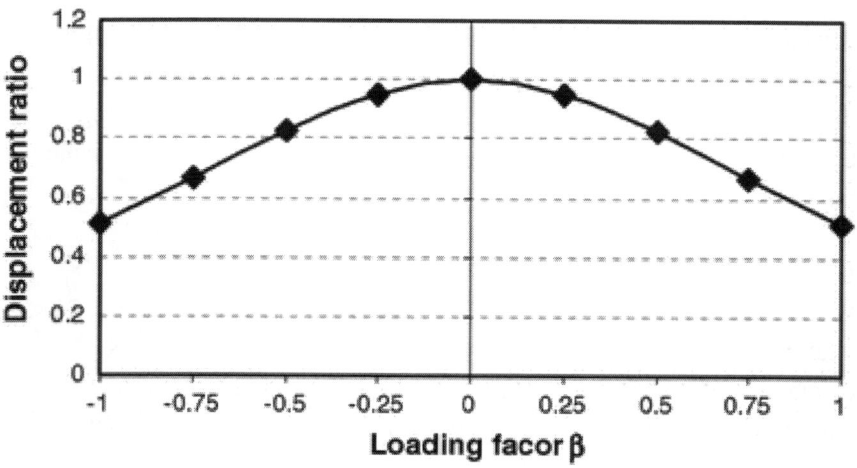

Figure 3. Displacement profile due to roller expansion

The boundary condition in this model was applied by constraining the primary side of tube from moving axially (Fig. 1c) and the outer side of the tube at its secondary side from moving tangentially. The model used in the current investigation did not require that the analysis includes the effects of torsional moments, because the tube was constrained from rotation. The local tangential strain was found to be insignificant compared to the global radial strain (Al-Aboodi 2006). This is due to the fact that the roller is sliding inside the tube as in Williams (1997) and Metzger et al. (1995). This reduces the amount of friction and tangential strain.

Similar to the work of Williams (1997) four rolling revolutions and two smoothing revolutions were used. No radial displacement is assigned during the smoothing process. To take into account the non-linear response of tube material in the elastic–plastic transition region and in the plastic zone the step size is determined using a non linear equation. The proposed relation yields a large step size in the elastic region and smaller and smaller sizes in the plastic region. The mathematical representation of such trend can be expressed in logarithmic profile as:

$$u_S = \frac{u_t \ln(n)}{N} \tag{3}$$

where usus is the step radial displacement, n is the step number, and N is the total number of steps.

2. RESULTS AND DISCUSSIONS

This section discusses the results on the deformation and stress distributions along the roller expanded length and transition zone in tube–tubesheet joint as well as the effect of large clearance values on the interfacial stresses. Figure 4 shows the tube inner and outer surface radial deformations along the tubesheet width at the end of the loading and unloading events. The radial displacement of the inner tube surface is indicated by u_{ri} and that of the outer surface by u_{ro}. Similar to what was reported in earlier work using 2-D axisymmetric model (Merah et al. 2010) inner and outer radial displacements for 5% wall reduction and an initial clearance of 0.127 mm remain constant in most of the 47.25 mm expanded length. A disturbance in the u_{ri} distribution happens in the transition zone (i.e. zone between the expanded and unexpanded areas of the tube). As will be explained later, this disturbance will have an effect on the stress distributions. Releasing the load caused the inner tube surface to spring back from a maximum of 0.232 to 0.188 mm resulting in radial strain of about 0.45% and that of the outer surface from 0.190 to 0.151 mm.

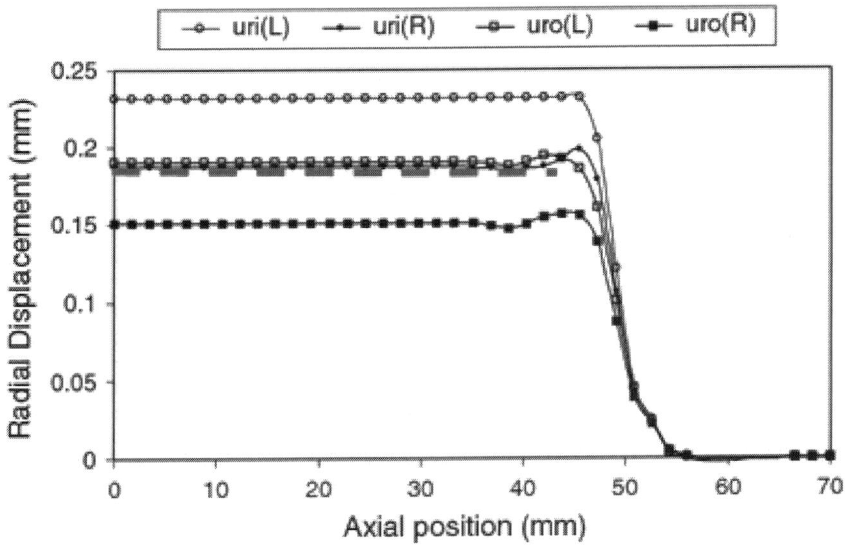

Figure 4. Loaded (L) and residual (R) radial displacement at the tube inner and outer surfaces (E_{tt} = 0.8 GPa, 5% WR, and c = 0.127 mm)

To experimentally validate these results tube diameters were measured before and after rolling. Several measurements were made for tube material having similar value of tangent modulus, E_{tt}, joint geometry and configuration, tube–tubesheet clearance, and loading condition (i.e. E_{tt} = 0.74 GPa, WR = 5% and$c$ = 0.12 mm). A micrometer was used to measure inner diameters of a number of tubes before and after expansion to determine the average value of the inner surface radial deformation. The thick dashed line in Fig. 4 shows the average value of the measured tube inner surface radial deformation (0.185 mm). It is clear that the present 3-D FE model yields an excellent representation of the deformed inner surface.

Figure 5 illustrates the distributions of stress components on the tube's inner and outer surfaces along the axial direction after the retraction of the rollers for the condition of 5% WR and 0.127 mm radial clearance. The stresses include σ_r, σ_θ, σ_y, σ_{eqv}, σ_{cont} which are the radial, tangential, axial, equivalent and contact stresses, respectively. The stresses corresponding to the tube's outer surface display a behavior similar to that of the radial displacement shown in Fig. 4. These stresses remain uniform along most of the expanded length and redistribute in the transition zone between the fully expanded and the unexpanded tube. At the end of the rolled tube length, the radial, axial and tangential stress components (σ_r, σ_z, σ_θ) exhibit two inflection points where they drop first and then increase very sharply in the transition zone. This behavior is attributed to the tube deformation elbow in this area (Fig. 4).

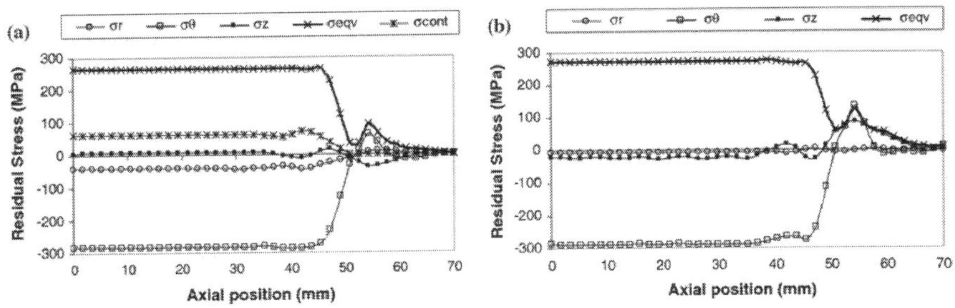

Figure 5. Stress distributions on the (**a**) outer and (**b**) inner tube surfaces after unloading (E_{tt} = 0.8 GPa, 5% WR, and$c$ = 0.127 mm)

Due to tubesheet spring back after releasing the load, the hoop stress (σ_θ) on both inner and outer surfaces is found to be compressive in most of the rolled tube length, but becomes tensile in the transition zone where it reaches a value of about half the yield strength of the material on the inner surface of the tube. The axial stress which remains very small in magnitude on both surfaces becomes tensile on the inner surface in the transition zone as well after roller retraction, reaching a value of 100 MPa. It should be noted that the tensile hoop and axial stress maxima happen at the same axial position around 8 mm from the end of the rolled area (Fig. 5b). This is somehow different from what was reported by Williams (1997). The existence of high tensile hoop and axial stresses in this region when coupled with expected tensile hoop stress from inner fluid pressure make it a critical design zone that needs proper attention especially when using aggressive fluids.

Analysis of the tube's outer surface stress distribution after roller retraction showed that σ_z reaches an average tensile stress value of 9 MPa which drops to 5 MPa upon the release of load which are very small compared to σ_r and σ_θ and emphasizes the assumption of this case as a plane stress model. The contact stress which reached an average compressive value of 160 MPa after the last loading step drops to 61 MPa upon unloading (Fig. 5a). Because this stress is a measure of the joint strength its variation with initial clearance and material strain hardening is studied in detail in the following section.

2.1 Effect of clearance and tube material strain hardening on residual contact stress

The combined effects of clearance and tube material strain hardening on the average value of radial stress for 5% wall reduction obtained from the 3-D finite element model are illustrated in Fig. 6. As can be seen, the clearance range considered in this study is well above the TEMA overtolerance limit. Similar to what was observed in axisymmetric model results by Al-Aboodi et al. (2008) for strain hardening tube materials ($E_{tt} > 0$ GPa), the residual contact stress gradually increases with the increase in radial clearance before dropping at lower cut-off clearances. The increase in residual contact stress with increase in E_{tt} is due to higher tube material strength resisting tubesheet material spring back.

As illustrated in Fig. 7, the 3-D model resulted in lower critical (cut-off) clearances for both types of tube materials ($E_{tt} = 0$ GPa and $E_{tt} = 1.2$ GPa) and higher contact stresses for strain hardening material than the axisymmetric model. This discrepancy comes from the difference between the calculated wall reduction and the actual one. The same value of input displacement will result in lower value of actual percent wall reduction for the 3-D model because of the extra tangential displacement in the 3-D model. Subsequently, the critical clearance will be higher for axisymmetric model.

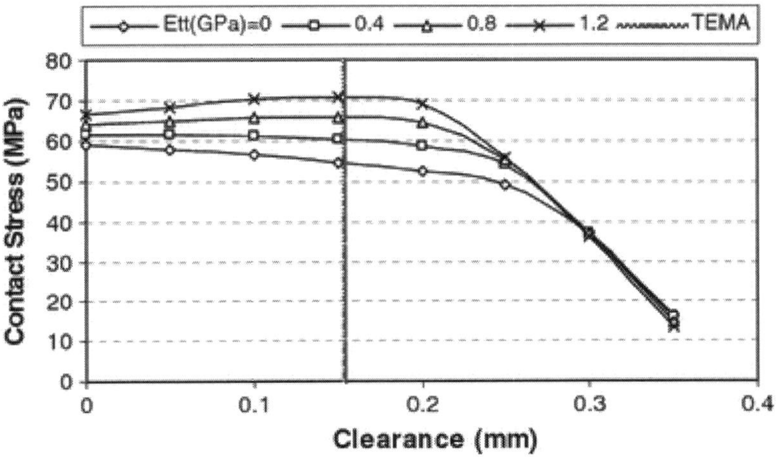

Figure 6. Effects of material strain hardening and clearance on contact stress for 5% WR

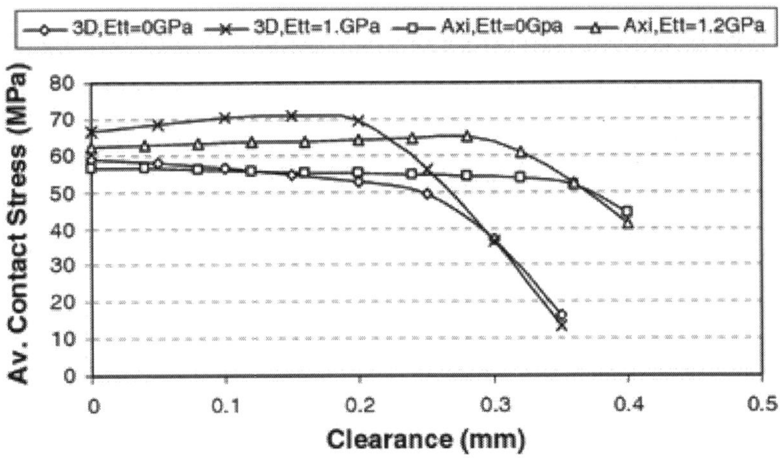

Figure 7. Comparison of axisymmetric and 3-D contact stress variation with clearance for different materials and for WR = 5%

2.2 Experimental validation of contact stress

In the experimental work the joint strength is measured in terms of pull-out or push-out force. The pull-out force F is directly proportional to the residual contact stress, σ_c, and the expanded length, l, (Scott et al.1984; Jawad et al. 1987).

$$F = 2\pi\mu\sigma_c r_o l \tag{4}$$

where r_o is the outer radius of the tube, and μ is the friction coefficient. The development of the above equation is based on the assumption that the expanded tube can be pulled or pushed out of the tubesheet by a force (F) equal to the friction force ($F_f = 2\pi\mu\sigma_c r_o l$) between the outer surface of the tube and the inner wall of the tubesheet hole. Two major challenges have to be dealt with when applying this equation to estimate the pull-out force; the friction coefficient and the contact stress. Different researchers have used different values for the friction coefficient. Metzger et al. (1995) used a value of 0.2. Allam and Bazergui (2002) developed an experimental set up and measured an average friction coefficient of 0.1375. The average value of the residual contact (radial) stress is usually used to estimate the pull out force (Merah et al. 2009; Merah 2008; Al-Aboodi et al. 2009; Shuaib et al. 2003; Al-Aboodi 2006). To validate the FE results, the pull out force F was computed using the contact stress determined by the 3-D model and $\mu = 0.1375$, for $E_{tt} = 0.8$ GPa and compared to that obtained experimentally by Shuaib et al. (2003) as well as to that determined from the axysmmetric model (Merah 2008). The material used in experimental program has an E_{tt} of 0.733 GPa. Using the empirical value of 0.1375, an outer tube radius of 9.525 mm, an expanded length of 47.25 mm the pull-out force is estimated to be between 23.65 and 26.5 kN within the TEMA (1988) clearance range. It can be seen from Fig. 8 that the pull-out force estimated from the 3-D model compares very well with the average value of the pull-out force obtained from the experimental tests (Shuaib et al.2003).

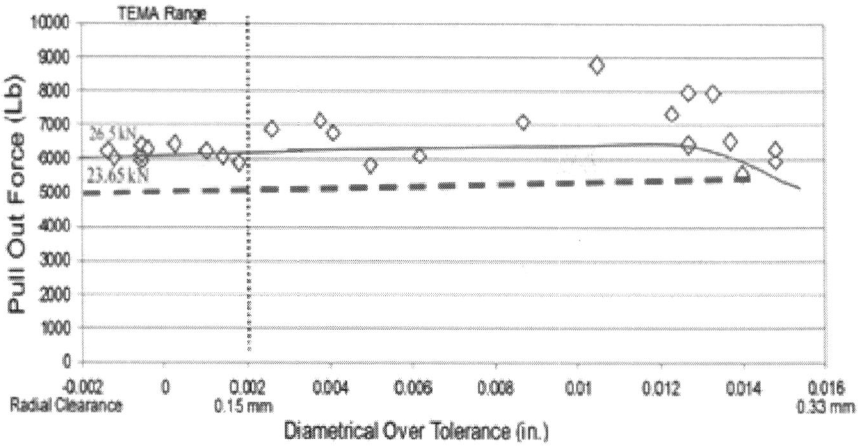

Figure 8. Comparison of pull-out force estimated by 3-D model to axisymmetric and experimental results

3. CONCLUSION

A 3-D finite element model is developed and utilized to investigate the radial displacement and stress distributions along the axial direction as well as the effect of radial clearance and tube material on the strength of a rolled tube–tubesheet stabilizer feed/bottom exchanger. The radial deformation distribution of inner and outer tube surfaces was uniform in most of the expanded tube length. A sudden increase in radial displacement is observed at the end of the expanded length (in the transition area). Similarly, inner tube surface residual axial and hoop stresses are found to be uniform and compressive along the expanded length and tensile in the transition zone between the expanded and the unexpanded tube regions. Both maximum tensile stresses happen at about 8 mm from the end of expanded length. The circumferential stress is predominant reaching a value around one-half the yield strength of the tube material. This, when combined with service load and corrosive environment is expected to cause longitudinal cracks.

The effect of clearance on residual interfacial pressure determined using the 3-D model is compared to that of axisymmetric FE model. Both models are found to yield comparable results showing that for elastic perfectly plastic tube material the residual contact pressure remains almost constant well above the prescribed TEMA maximum overtolerance values. In addition, both models show that for strain hardening tube materials the interfacial pressure increases with increasing clearance; with 3-D model predicting higher radial stress values. An appreciable difference is observed at high overtolerances where the 3-D model predicts cut-off clearances (clearance at which the interfacial pressure starts to drastically drop) which are about 30% lower than those predicted by the axisymmetric model. The experimental validation proved that the present 3-D model allowed a more realistic estimation of the inner tube wall deformation and pull out force.

ACKNOWLEDGMENTS

The authors thankfully acknowledge the support of the King Fahd University of Petroleum and Minerals. The second author acknowledges the support of Qassim University.

REFERENCES

1. Al-Aboodi A (2006) Finite element analysis of roller expanded tube-tubesheet joints with overenlarged tubesheet holes. PhD thesis, KFUPM

2. Al-Aboodi A, Merah N, Shuaib AR, Al-Nassar Y, Al-Anizi SS (2008) Modeling the effects of initial tube-tubesheet clearance, wall reduction and material strain hardening on rolled joint strength. ASME J Press Vessel Technol 130/4:041204-1–041204-6

3. Al-Aboodi A, Merah N, Shuaib AN, Al-Nassar Y (2009) FEA of groove geometry effect on roller and hydraulically expanded tube to tubesheet joint strength. Int J Precis Technol 1(2):201–207

4. Allam M, Bazergui A (2002) Axial strength of tube-to tubesheet joints: finite element and experimental evaluations. J Press Vessel Technol 124:23–31

5. Allam M, Chaaban A, Bazergui A (1998) Estimation of residual stresses in hydraulically expanded tube-to-tubesheet joints. J Press Vessel Technol 120:129–137

6. Al-Zayer A (2001) Analytical and finite element investigation of initial clearance effect on tube-to-tubesheet joint strength. MSc thesis, KFUPM

7. Andrieux S, Voldoire F (1995) Stress identification in steam generator tubes from profile measurements. Nucl Eng Design 158:417–427

8. ANSYS (2004) Version 9.0, program and help documentations, Swanson Analysis System, Inc.

9. Aufaure M (1987) Analysis of residual stresses due to roll-expansion process: finite element computation and validation by experimental tests. In: Transaction of the 9th international conference of SMIRT, pp 499–503

10. Cizelj L, Mavko B (1995) Propagation of stress corrosion cracks in steam generator tubes. Int J Press Vessel Piping 63:35–43

11. Cooper Power Tools (2005) Tube cleaners and expanders airetool® manual. SP-1100EN0405–410M

12. Jawad MH, Clarkin EJ, Schuessler RE (1987) Evaluation of tube-to-tubesheet junctions. J Press Vessel Technol 109:19–26

13. Merah N, Al-Zayer A, Shuaib A, Arif A (2003) Finite Element evaluation of clearance effect on tube-to-tubesheet joint strength. Int J Press Vessel Piping 80:879–885

14. Merah N, Al-Aboodi A, Shuaib AN, Al-Nassar Y, Al-Anizi SS (2009) Combined effects of tube projection, initial tube-tubesheet clearance and tube material strain hardening on rolled joint strength. ASME J Press Vessel Technol 131/5

15. Merah N, Al-Aboodi A, Shuaib AN, Al-Nassar Y (2010) 3-D FEA of the effects of large overtolerances on roller expanded tube-tubesheet joint strength. In: Abstract published on IVth European conference on computational mechanics, Paris 16–21, 2010

16. Metzger DR, Sauve RG, Nadeau E (1995) Prediction of residual stress by simulation of the rolled joint manufacturing process for steam generators, PVP vol 305. In: Current topics in computational mechanics. ASME, New York

17. Scott DA, Wolgemuth GA, Aikin JA (1984) Hydraulically expanded tube-to-tubesheet joints. J Press Vessel Technol 106:104–109

18. Shuaib AN, Merah N, Khraisheh MK, Allam IM, Al-Anizi SS (2003) Experimental investigation of heat exchanger tubesheet hole enlargement. J Press Vessel Technol 125:19–25

19. Standard of the Tubular Exchanger Manufacturer Association 'TEMA', 7th edn (1988) TEMA, Terrytown

20. Updike DP, Kalnins A, Caldwell SM (1992) Residual stresses in transition zone of heat exchanger tubes. J Press Vessel Technol 114:149–156

21. Williams DK (1996) Prediction of residual stresses in the mechanically expanded 0.750″ diameter steam generator tube plugs—part 1: 2-D solution, PVP vol 327. In: Residual stresses in design, fabrication, assessment and repair, ASME, New York, pp 173–180

22. Williams DK (1997) Prediction of residual stresses in the mechanically expanded 0.750″ diameter steam generator tube plugs—part 2: 3-D solution, PVP vol 354. In: Current topics in the design and analysis of pressure vessels and piping. ASME, New York, pp 17–28

23. Williams DK (2003) Prediction of residual stresses in the mechanical roll of HX tubes into TEMA grooves, PVP vol 2003–1937. In: Design and analysis methods and fitness for service, ASME, New York, pp 121–29

Index

H

L

M

N

O

P

S

T

V

Z